QUALITATIVE DATA ANALYSIS WITH CHATGPT AND QUALCODER

A Step-By-Step Guide to AI-Powered Coding and Thematic Analysis

QUALITATIVE DATA ANALYSIS WITH CHATGPT AND QUALCODER

A STEP-BY-STEP GUIDE TO AI-POWERED CODING AND THEMATIC ANALYSIS

RAFIQ MUHAMMAD, MD, MIHMEP, PH.D.

Disclaimer:
No part of this book is to be duplicated, shared, or transmitted in any manner without explicit written authorization from the author. This prohibition extends to all forms of reproduction, whether by copying, recording, or using any mechanical or electronic means, as well as storing or retrieving information through any system. While certain segments of this book have benefitted from advanced AI tools, extensive human intervention in the form of editing and revising has been undertaken to ensure its precision, clarity, and reliability.

Notice of Liability:
It is vital to acknowledge that the information provided in this book comes with no explicit guarantees. The author cannot be held accountable for any losses or damages that may arise from the content or links contained in this text.

Websites and links
Given the ever-evolving nature of the internet, websites and their content are subject to continual changes. The hyperlinks referenced in this book are intended solely for informational purposes. The author does not warrant the accuracy, reliability, or any other aspects of the content found on external websites linked within this book.

ISBN: 978-91-989008-4-2

Imprint: Muhammad Rafiq

©Copyright Rafiq Muhammad 2024, Protected by Copyright Law

TABLE OF CONTENTS

TABLE OF CONTENTS 5
WHO SHOULD READ AND WHAT YOU WILL GAIN 9
    WHAT YOU WILL LEARN 12
ABOUT THE AUTHOR 14
INTRODUCTION 15
    WHY THIS BOOK? 19
    WHAT IS INCLUDED IN THIS BOOK 21
PART I: FOUNDATIONS OF LARGE LANGUAGE MODEL IN QUALITATIVE ANALYSIS 23
UNDERSTANDING CHATGPT AND AI LANGUAGE MODELS 24
THE ROLE OF AI IN QUALITATIVE RESEARCH 32
    Historical Context of Qualitative Methods 32
    Debunking Myths About AI Replacing Researchers 35
PART II: PRACTICAL STRATEGIES FOR USING CHATGPT 39
INDUCTIVE AND DEDUCTIVE APPROACHES WITH CHATGPT 40
    Approaches to Qualitative Data Analysis with ChatGPT: Inductive and Deductive Approaches 40
    Inductive Approach to Qualitative Data Analysis 40
    Deductive Approach to Qualitative Data Analysis 43
    Inductive vs. Deductive: Which Approach to Use? 44
    Limitations and Considerations for Using ChatGPT 45
    Choosing the Right Approach 46
EFFECTIVE PROMPT ENGINEERING 50
    Crafting Prompts for Optimal Responses 50
    Techniques for Iterative Questioning 54
    Handling Ambiguous or Complex Data Inputs 56
AI-ASSISTED OPEN CODING 60
    AI-Assisted Open Coding 60
    What Is Open Coding? 60
    Balancing AI Suggestions with Researcher Intuition 62

Documenting the Coding Process for Transparency 64
Setting the Stage for AI-Assisted Open Coding 65

QUALCODER: A FREE, OPEN-SOURCE TOOL FOR QUALITATIVE ANALYSIS 67

Introduction to QualCoder 67
Overview of QualCoder's Features 68
Why Choose Open-Source Software for Qualitative Analysis 69
Getting Started with QualCoder 72
Importing Data into QualCoder 75
Organizing and Managing Data within Projects 78
Coding and Categorization in QualCoder 80
Analyzing Data in QualCoder 85

INTEGRATING CHATGPT WITH QUALCODER 91

Manual import 94
Semi-Automated Transfer Using CSV or TXT Files 99
Fully Automated Import via API Integration 104

PREPARING YOUR RESEARCH ENVIRONMENT FOR AI INTEGRATION 112

Accessing and Configuring ChatGPT 112
Choosing the Right Approach 115
Data Preparation: Cleaning and Formatting Qualitative Data 118
Transcription Accuracy 118
Formatting for AI Readability 118
Breaking Down Large Data Sets 119
Data Cleaning for Themes 120
Example Workflow for Data Cleaning: 121
Anonymizing Data 123
Informed Consent 124
Compliance with Data Protection Laws 124
Storage and Data Security 125
OpenAI's Data Usage Policies 127
Transparency and Ethical Reporting 127
Interview transcript example 129
Deductive Thematic Analysis 129
Structure of the Interview 130
Predefined Themes for Deductive Analysis 132
Hypothetical Interview Transcript 134
Safeguarding Privacy in AI-Assisted Research 138

STEP-BY-STEP QUALITATIVE DATA ANALYSIS WITH CHATGPT AND QUALCODER 139

Step 1: Clean the Interview Transcript 140

Step 2: Set Up ChatGPT for Qualitative Data Analysis 142
Step 3: Define Themes and Develop Prompts for Deductive or Inductive Analysis 144
Step 4: Create Prompts and Run the Coding Process 151
Step 5: Import Data into QualCoder and Generate a Code Tree 156
Step 6: Analyze Codes in QualCoder 166
Step 7: Develop Themes and Categories 167
Step 8: Thematic Analysis in ChatGPT 168
Step 9: Validate and Cross-Check 172
Step 10: Visualize and Analyze in QualCoder 173

PART III: CREDIBILITY ENHANCEMENT AND REPORTING WITH AI 178

ENHANCING CREDIBILITY AND RELIABILITY 179

Triangulating AI Findings with Traditional Methods 179
Strategies for Mitigating AI Biases 181
Peer Debriefing and Member Checking with AI Assistance 182

REPORTING AND PRESENTING FINDINGS 185

Crafting Compelling Reports Using AI Insights 185
Ethical Considerations in Attributing AI Contributions 187
Communicating Complex Themes to Diverse Audiences 188
ONE LAST THING 192

REFERENCES 193

Are you just starting your research journey and feeling overwhelmed?

Get my FREE 75-page booklet, **Getting Started with Research Design: A Quick Guide to Qualitative, Quantitative, and Mixed Methods Research**.

Inside this booklet, you will find practical tips, checklists, and clear explanations to help you confidently craft your research question, select the right methodology, plan your data collection, and get them published.

Get the foundation you need to kickstart your research with confidence—click the **LINK to download** your free copy now!

Who Should Read and What You Will Gain

You sit at your desk, surrounded by interview transcripts, notes, and a blinking cursor on your computer screen. The challenge before you is clear: make sense of the vast amount of qualitative data you have collected. How do you distill meaning from these stories? How do you organize your codes and themes in a way that stays true to the voices of your participants? You lean back and think, "There has to be a better way to streamline this process."

As you begin typing your first set of codes into a spreadsheet, the thought lingers in your mind: "Can I possibly capture all the nuance here?" Your hand hovers over the keyboard, contemplating if there are themes you have not yet considered. "Am I missing something important?" As the volume of data grows, the task ahead seems daunting. Hours of manual coding and thematic analysis are stretching ahead of you, and yet you wonder: is there a more efficient way to achieve accuracy without compromising depth?

This scenario is familiar to many qualitative researchers, particularly those working through the coding and thematic analysis phases of their studies. The sheer volume of data, coupled with the need for rigorous analysis, can often feel overwhelming. Without clear strategies or tools, researchers can find themselves lost in the depths of data management, struggling to synthesize key themes while maintaining the richness of the data.

At that moment, an idea crosses your mind: "Could Artificial Intelligence (AI) help me with this?" You have heard about AI tools like ChatGPT being used to assist with tasks that involve language, but could it really be useful in something as complex as qualitative data analysis? You pause, feeling both excitement and hesitation, unsure if AI can maintain the integrity of your research or if it is just another overhyped tool.

But curiosity drives you forward.

You open a new document and type in a test prompt, asking ChatGPT to assist in summarizing a segment of your interview transcript. A few moments later, you are impressed by the clarity and relevance of the response.

"This could really speed things up," you think, as the path to integrating AI into your research becomes clearer.

The biggest hurdles in qualitative research often involve labor-intensive tasks such as coding data, generating themes, and making sense of complex narratives. These tasks can be time-consuming and overwhelming, especially when balanced with the need for methodological rigor. This book breaks down these challenges into manageable steps, guiding you through the process of using ChatGPT and QualCoder together as valuable tools in qualitative analysis.

At the same time, you recognize a challenge faced by many students and early-career researchers: the high cost of traditional qualitative data analysis software, like MAXQDA, ATLAS.ti, and NVivo, often makes these tools inaccessible.

This book aims to overcome these barriers by offering a comprehensive, step-by-step guide on using ChatGPT for qualitative data analysis, while integrating it with QualCoder, a widely recognized, free, and open-source qualitative data analysis software.

Whether you are a seasoned qualitative researcher or just starting your journey, this guide will show you how to leverage the combined power of ChatGPT and QualCoder to streamline and enhance your qualitative analysis process. By integrating ChatGPT's AI capabilities with the open-source flexibility of QualCoder, you can work more efficiently without sacrificing the depth and rigor required in qualitative research. This dynamic combination allows you to focus more on interpretation and insights, while the tools assist with coding, pattern recognition, and thematic development.

The focus of this book is to demystify AI in qualitative research, emphasizing the integration of ChatGPT with QualCoder. You will learn how to set up and configure these tools, craft effective prompts for coding and thematic analysis, and merge AI-generated insights with your expertise. From data preparation to reporting, this guide covers essential strategies to help you maintain control over your research process while benefiting from AI-driven efficiency.

This book is ideal for those who:

- Are new to qualitative research and want to explore how AI and open-source tools like QualCoder can assist in coding and analysis.
- Have experience with qualitative methods and are looking to boost productivity through ChatGPT and QualCoder.
- Are interested in responsibly incorporating AI without compromising research integrity.

By the end, you will have the skills and insights to streamline coding, develop themes with AI support, and gain deeper insights from your data. You will also navigate the ethical considerations involved, ensuring your research remains transparent, reliable, and valid. Whether you are overwhelmed by coding, unsure about structuring your analysis, or curious about AI's potential, this book provides the guidance you need to make impactful findings from your data.

WHAT YOU WILL LEARN

Here is what you will learn in this book:

- **Foundations of AI-Powered Qualitative Research**: Gain a solid understanding of how AI, particularly ChatGPT, can be integrated into qualitative research, including the benefits and limitations of using AI for coding and thematic analysis. You will also learn how to combine ChatGPT with QualCoder, a free, open-source qualitative analysis tool, to enhance your research process without the need for expensive software.
- **Crafting Effective Prompts for Qualitative Analysis**: Learn the art of designing clear and precise prompts for ChatGPT that elicit meaningful responses, helping you streamline tasks such as open coding, pattern recognition, and data summarization. You will also see how these prompts work within QualCoder to support a more structured analysis.
- **Data Preparation and Privacy Considerations**: Discover essential techniques for cleaning, organizing, and formatting your qualitative data to optimize ChatGPT's performance while ensuring that privacy and ethical standards are maintained throughout your research. Additionally, learn how to prepare your data for seamless integration with QualCoder for efficient analysis.
- **AI-Assisted Open Coding and Thematic Development**: Understand how to collaborate with ChatGPT during the coding process to generate initial codes and identify emerging themes. You will also learn how to use QualCoder to refine these codes and ensure your analysis maintains research integrity, blending AI suggestions with your own insights.
- **Integrating ChatGPT with QualCoder**: Explore practical strategies for combining ChatGPT with QualCoder, a powerful, free, open-source software for qualitative data analysis. Learn how to leverage the strengths of both tools to streamline coding, pattern recognition, and thematic development, making high-quality qualitative analysis more accessible.
- **Advanced AI Techniques for Data Interpretation**: Learn how to use ChatGPT to dive deeper into sub-themes, compare different data sets, and synthesize complex findings into coherent narratives. You

will also discover how to integrate these insights into QualCoder for more comprehensive data management and visualization.

- **Ensuring Credibility and Mitigating Bias**: Understand how to validate AI-generated insights, triangulate findings with traditional methods, and implement strategies to reduce AI biases, ensuring reliable and ethical research outcomes. By using ChatGPT alongside QualCoder, you will be able to maintain credibility and ensure a balanced analysis.

While the focus of this book is on using ChatGPT as a powerful AI tool for qualitative research, it is important to recognize that there are several other large language models (LLMs) available that can be used similarly. Tools such as Google's Gemini, Anthropic's Claude, and Cohere provide advanced language processing capabilities, much like ChatGPT. Each of these models offers unique features, but the core principles of integrating them into your research process—such as prompt crafting, coding assistance, and thematic development—remain consistent. As you progress through the book, keep in mind that the techniques discussed can be adapted to whichever LLM best suits your needs or resources.

Additionally, while many qualitative researchers turn to paid software like NVivo, MAXQDA, or ATLAS.ti for their data analysis, this book focuses on integrating ChatGPT with QualCoder, a free, open-source alternative. QualCoder provides many of the same features as commercial tools—such as coding, categorizing, and managing qualitative data—without the financial barriers. By demonstrating how to combine QualCoder with ChatGPT for tasks like automated coding, pattern identification, and theme generation, this book ensures that even researchers with limited budgets can benefit from the power of AI in their qualitative analysis.

ABOUT THE AUTHOR

Rafiq Muhammad, MD, MIHMEP, Ph.D., is a seasoned academic and medical professional with a rich background in healthcare research. He earned his medical degree followed by a Master of International Healthcare Management, Economics and Policy from the prestigious SDA Bocconi School of Management in Italy, and a Ph.D. from the world-renowned Karolinska Institute in Sweden.

With a robust track record in the academic publications, Dr. Rafiq has contributed extensively to peer-reviewed journals, enhancing the body of knowledge in medical research. His expertise extends beyond writing to teaching and mentoring graduate students. With a strong foundation in research design, Dr. Rafiq has contributed to the academic community through numerous peer-reviewed publications and has successfully guided students in navigating their own research journeys. This book distills that expertise, offering practical strategies and insights to help you confidently design and conduct your research.

In "Qualitative Data Analysis with ChatGPT," Rafiq draws on personal experiences and professional expertise to equip researchers with the tools needed to excel in their qualitative research projects. Whether you are exploring AI-assisted analysis for the first time or seeking to enhance your current research skills, this book offers the clarity, confidence, and practical knowledge necessary to succeed in an ever-evolving research landscape.

Chapter 1.

INTRODUCTION

Imagine This!

You find yourself in the middle of an exciting academic discussion, where the focus is shifting to the use of AI—specifically ChatGPT—in qualitative research. The prospect of integrating AI into the field is gaining momentum, with many recognizing the transformative potential it offers for navigating large, complex data sets. Researchers are beginning to see how ChatGPT can help streamline the traditionally time-consuming processes of coding, identifying themes, and making sense of narrative data (1–3). This enthusiasm, however, comes with questions: Can AI truly assist with qualitative analysis? How can it be integrated with established qualitative methods while maintaining the depth and rigor the field demands?

As you dive deeper into these discussions, you come across an article titled *"A New Era for Data Analysis in Qualitative Research: ChatGPT!,"* which showcases how the tool's ability to process massive amounts of textual data can dramatically speed up tasks like coding and theme generation (2). On the other hand, a colleague points out that the use of AI in qualitative research is still relatively new and evolving, with many researchers just beginning to explore how it fits into their workflows (4). This evolving landscape opens up a new conversation: how can ChatGPT, as an emerging tool, address the core challenges faced in qualitative research—namely, managing complex data sets, integrating technology with traditional methods, and presenting data in an engaging and meaningful way?

Your own research experiences may reflect this dynamic. Perhaps you are faced with a mountain of interview transcripts or field notes, wondering how to effectively organize and analyze this data. The challenge of navigating large, unstructured datasets is one of the biggest hurdles in qualitative research (5). ChatGPT offers a solution by acting as a powerful assistant—helping you manage, sort, and highlight key patterns in your data. But how do you ensure that this AI-driven analysis complements your own interpretations? How can you harness ChatGPT's capabilities to enrich, rather than overshadow, the depth of your qualitative insights?

How ChatGPT Addresses Key Challenges in Qualitative Analysis

One of the main strengths of ChatGPT is its ability to handle complex and voluminous qualitative data with ease. Imagine trying to manually sift through hundreds of pages of interview transcripts, each filled with rich and detailed narratives. This can be an overwhelming task, but ChatGPT can help by quickly organizing and identifying recurring themes, phrases, or concepts across the data set.

This does not replace your work as a researcher—it enhances it.

ChatGPT allows you to navigate the data more efficiently, freeing up time for the deeper analysis that only human insight can provide.

For example, ChatGPT can assist by generating preliminary codes and suggesting patterns, but it is up to you to refine these findings and interpret their meaning within the context of your study. The integration of AI with traditional qualitative methods means that you can blend technology's efficiency with your interpretive skills, ensuring that the richness of the data is preserved. This book will guide you through the process of using ChatGPT to navigate complex data while maintaining the rigor and depth that qualitative research demands.

Integrating ChatGPT with Traditional Qualitative Methods

While the use of AI in qualitative research is still emerging, ChatGPT is a powerful tool that seamlessly complements traditional methods. One of the key challenges in qualitative research is balancing the need for technological assistance with the human-driven process of interpretation. ChatGPT excels at managing large amounts of text and identifying broad patterns, but the true value comes when this technology is used alongside traditional techniques such as coding by hand, thematic analysis, and narrative synthesis.

By using ChatGPT to handle initial stages—such as sorting data, suggesting codes, and identifying high-level themes—you can focus on the more nuanced aspects of interpretation. This allows for a more dynamic research process where technology supports your expertise rather than replacing it. Throughout this book, you will learn how to incorporate ChatGPT into each step of your analysis, ensuring that the tool enhances your work without compromising the integrity of your research.

Presenting Data in an Engaging Manner with ChatGPT

Another challenge in qualitative research is presenting your findings in a way that is both accurate and engaging. The richness of qualitative data comes from its ability to tell stories, and ChatGPT can be an invaluable ally in this process. By using AI to organize and structure your data, you can more easily identify the key insights that will form the foundation of your narrative. ChatGPT's ability to summarize and synthesize complex information can help streamline the presentation process, making your findings clearer and more accessible to your audience.

However, while ChatGPT can help structure the data, the heart of qualitative research lies in the interpretation and storytelling. This book will show you how to use ChatGPT's capabilities to present your data in a way that maintains the depth and nuance of the original narratives, while also ensuring that your findings are clear, concise, and impactful. You will learn how to take AI-generated themes and weave them into a compelling narrative that captures the complexity of human experiences.

Why This Book Focuses on ChatGPT for Qualitative Research

In a rapidly evolving research landscape, ChatGPT offers a unique opportunity to address some of the most pressing challenges in qualitative analysis. This book is designed to help you navigate the complex world of qualitative research using ChatGPT as a powerful tool to enhance your analysis. From managing large data sets to integrating AI with traditional methods, and presenting your findings in an engaging way, this book provides practical guidance on how to use ChatGPT to transform your qualitative research process.

Whether you are a seasoned researcher looking to incorporate new technologies or a student just beginning to explore the possibilities of AI, this book will equip you with the knowledge and skills to effectively use ChatGPT in your qualitative research. You will learn how to make the most of AI's capabilities while ensuring that your research remains grounded in the interpretive depth and rigor that qualitative analysis demands.

WHY THIS BOOK?

In creating this guide, I have thoughtfully crafted each chapter with you in mind—whether you are an undergraduate just starting out, a graduate student delving deeper, or a seasoned researcher navigating the innovative landscape of qualitative data analysis with ChatGPT. My goal is to simplify the complexities of integrating AI into your research and to provide you with the tools and confidence needed to embrace this transformative approach.

What Sets This Book Apart:

- **Actionable Steps:** Break down the integration of ChatGPT into your qualitative analysis with clear, practical tips that alleviate any fears about using AI in your research.
- **Practical Solutions for Modern Challenges**: This book tackles the real-world issues researchers face in qualitative analysis, such as managing large, complex datasets and combining traditional methods with AI technologies like ChatGPT. You will find strategies that help you navigate these challenges efficiently, making your research process smoother and more effective.
- **Focus on Open-Source and Affordable Tools:** Discover QualCoder, a robust, open-source, and free tool for qualitative data analysis, allowing you to perform in-depth analysis without the high costs associated with other software.
- **Comprehensive Coverage:** From understanding the basics of ChatGPT and AI-assisted coding to presenting your findings with AI-enhanced analysis, this guide covers the entire qualitative research process in one cohesive resource.
- **Clear Explanations:** Complex AI and natural language processing concepts are explained in straightforward language, complemented by real-world examples that make it easy to apply these ideas effectively.
- **For All Levels:** Whether you are new to qualitative research or have extensive experience, this book addresses the unique challenges of adopting new technologies, offering tailored advice to help you overcome any hurdles with ChatGPT.

- **Valuable Resources:** Access templates, sample prompts, and insightful recommendations designed to enhance your research projects and seamlessly integrate ChatGPT into your qualitative analysis.
- **Organized Tools:** Utilize checklists and practical instruments to stay organized and ensure you do not miss any critical steps as you incorporate ChatGPT into your workflow.
- **Real-Life Case Studies**: The book features case studies and real-world examples of how AI has been successfully integrated into qualitative research projects. These practical applications show you exactly how to adapt the strategies discussed in this book to your own work, making abstract concepts concrete and actionable.

This book is more than just a guide—it is your reliable companion, empowering you to harness ChatGPT in your qualitative research. Whether you are embarking on your first project or refining your methods, you will find valuable insights and practical tools to make your research more efficient and impactful.

By the end of this book, you will have a clear understanding of how to use ChatGPT to streamline your qualitative research while maintaining the integrity and depth of your data. Whether you are managing a large dataset, looking for advanced AI integration techniques, or simply seeking to understand the basics, this book offers everything you need to confidently enhance your qualitative research with AI.

WHAT IS INCLUDED IN THIS BOOK

The book begins by laying a solid foundation for understanding ChatGPT and its role in qualitative research. In **"Understanding ChatGPT and AI Language Models,"** the basics of AI language models are explained, along with how ChatGPT can assist in qualitative data analysis. **"The Role of AI in Qualitative Research"** delves into the evolving role of AI, addressing concerns and possibilities for integrating AI with human interpretation. **"Inductive and Deductive Approaches with ChatGPT"** helps you understand how to apply these methodologies using AI tools.

As the book moves into more practical aspects, **"Effective Prompt Engineering"** guides you through crafting prompts to elicit useful and accurate responses from ChatGPT—a key skill for effective AI-assisted research. **"AI-Assisted Open Coding"** demonstrates how to use ChatGPT to assist with open coding, balancing AI-generated codes with your own interpretive insights and ensuring transparency in your coding process.

The book then introduces **"QualCoder: A Free, Open-Source Tool for Qualitative Analysis,"** providing an overview of this powerful software and instructions on how to get started with it. In **"Integrating ChatGPT with QualCoder,"** it explains how to combine ChatGPT with QualCoder.

"Preparing Your Research Environment for AI Integration" covers essential steps for setting up your research environment, including accessing and configuring ChatGPT, data preparation, and ensuring data privacy and compliance. This sets the stage for **"Step-by-Step Qualitative Data Analysis with ChatGPT and QualCoder,"** which provides a detailed walkthrough of the entire process—from cleaning your data to analyzing and visualizing results.

In the later chapters, **"Enhancing Credibility and Reliability"** focuses on strategies for mitigating AI biases, triangulating AI findings with traditional methods, and ensuring the credibility of your research. **"Reporting and Presenting Findings"** offers guidance on crafting compelling reports using AI insights, ethical considerations in attributing AI contributions, and effectively communicating complex themes to diverse audiences.

By the end of this book, you will have a comprehensive understanding of how to integrate ChatGPT and QualCoder into your qualitative research process—helping you manage large data sets, combine AI with traditional methods, and present your findings in a clear, engaging manner. Whether you are working on a thesis, dissertation, or professional research project, this book provides the knowledge, tools, and strategies to conduct high-quality research that leverages the power of AI while preserving the depth and interpretive nature of qualitative analysis.

Chapter Key Points:

- *Managing large, complex qualitative datasets is a significant challenge for researchers.*
- *Balancing AI integration with human interpretation without losing depth is crucial.*
- *Presenting qualitative findings in an engaging and meaningful way is often difficult.*
- *The book introduces ChatGPT as a transformative tool to address these challenges.*
- *It offers practical guidance on integrating ChatGPT with traditional qualitative methods.*
- *Enhancing efficiency with AI without sacrificing interpretive depth is a central focus.*
- *The book empowers researchers at all levels to confidently use AI tools like ChatGPT.*
- *Strategies are provided to mitigate AI biases and maintain scientific rigor.*
- *Ethical considerations and data privacy are addressed to ensure research integrity.*
- *The goal is to enhance qualitative research using ChatGPT, making projects methodologically sound and innovative.*

Part I: Foundations of Large Language Model in Qualitative Analysis

Chapter 2.

UNDERSTANDING CHATGPT AND AI LANGUAGE MODELS

To use ChatGPT to its full potential in your research, it is essential to first understand the basics of how AI and natural language processing (NLP) work, the capabilities and limitations of ChatGPT, and the ethical considerations that come with incorporating AI into your research process. By the end of this chapter, you will have a solid grasp of the mechanics of ChatGPT and be prepared to use it responsibly and effectively.

Basics of AI and Natural Language Processing

Artificial intelligence has been rapidly advancing in recent years, with NLP emerging as one of its most transformative branches. NLP allows machines to understand, interpret, and generate human language in ways that were previously impossible. At its core, NLP bridges the gap between human communication and computer comprehension, enabling AI to process and analyze vast amounts of unstructured text data—exactly the kind of data generated in qualitative research.

In the context of ChatGPT, NLP enables the AI to "read" through qualitative data, such as interview transcripts or focus group discussions, and assist in identifying patterns, codes, and themes. Key techniques in NLP include:

- **Tokenization**: Breaking down text into smaller units like words or phrases.
- **Text classification**: Assigning predefined categories to text, which can assist in tasks like sentiment analysis or thematic coding.
- **Named entity recognition**: Identifying and classifying specific elements like names, dates, or organizations within a text.

- **Text generation**: The ability of AI, like ChatGPT, to generate human-like responses or summaries from prompts or questions.

ChatGPT, specifically, is built on the GPT (Generative Pretrained Transformer) architecture, which uses deep learning models to predict the next word in a sequence, making it highly proficient at tasks like conversation generation and text summarization (6).

In qualitative research, these features can streamline tasks like coding and thematic analysis, saving researchers time and energy.

How ChatGPT Works: Capabilities and Limitations

To understand how ChatGPT functions in a research setting, let us break down its capabilities and limitations (Figure 2.1).

Capabilities:

1. **Text Summarization**: ChatGPT can process large amounts of text data and generate concise summaries, helping researchers quickly identify the core ideas or themes from lengthy transcripts.
2. **Text Classification and Coding**: ChatGPT can assist with the open coding process by identifying recurring words, phrases, or concepts in qualitative data. For example, if you input raw interview transcripts, ChatGPT can highlight key terms and suggest initial codes, which you can refine and validate.
3. **Generating Ideas and Questions**: ChatGPT is useful for brainstorming, generating questions, or expanding on thematic insights. This can be particularly helpful in refining research questions or developing new angles of inquiry during thematic analysis.
4. **Pattern Recognition**: By analyzing qualitative data for trends and recurring themes, ChatGPT can support pattern identification, helping you find connections in the data you might otherwise overlook.

Figure 2.1: Capabilities of ChatGPT

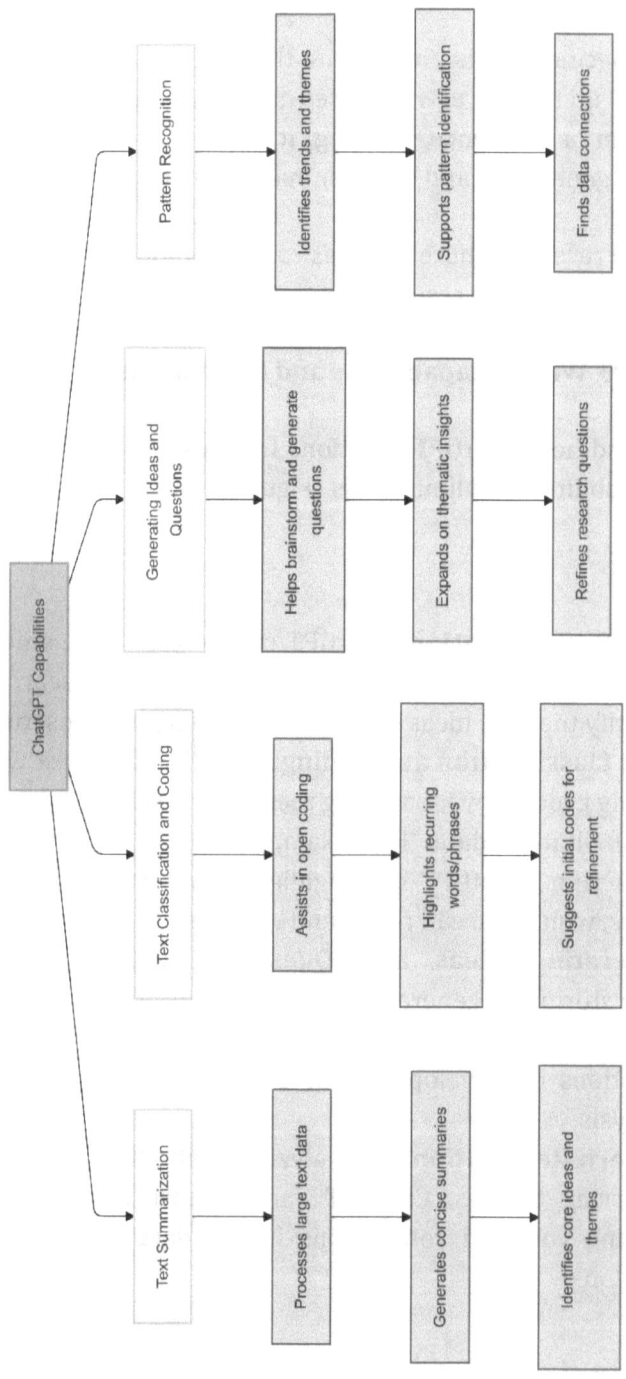

However, while ChatGPT has impressive capabilities, it is important to recognize its limitations (Figure 2.2).

Limitations:

1. **Contextual Understanding**: Although ChatGPT can generate coherent text, it does not *truly* understand the context in which the data is being analyzed. For instance, while it may suggest codes based on language patterns, it cannot fully comprehend cultural nuances, emotional undertones, or social context—factors that are critical in qualitative research.
2. **Bias in Data**: ChatGPT is trained on vast datasets pulled from the internet, and as a result, it may reflect certain biases present in its training data. This can potentially influence the codes or themes it generates, necessitating careful oversight from the researcher to ensure accuracy and objectivity.
3. **Inconsistent Quality**: While ChatGPT can produce highly relevant insights in one instance, it may generate less useful or even nonsensical responses in others. This variability requires researchers to cross-check AI-generated results against their own analysis to ensure validity.
4. **Ethical and Interpretive Challenges**: ChatGPT can assist in identifying patterns and suggesting themes, but it cannot interpret findings or offer theoretical insights. The interpretation of data remains a uniquely human task, grounded in your understanding of the research context and the theoretical framework guiding your study.

Figure 2.2. Limitations of ChatGPT

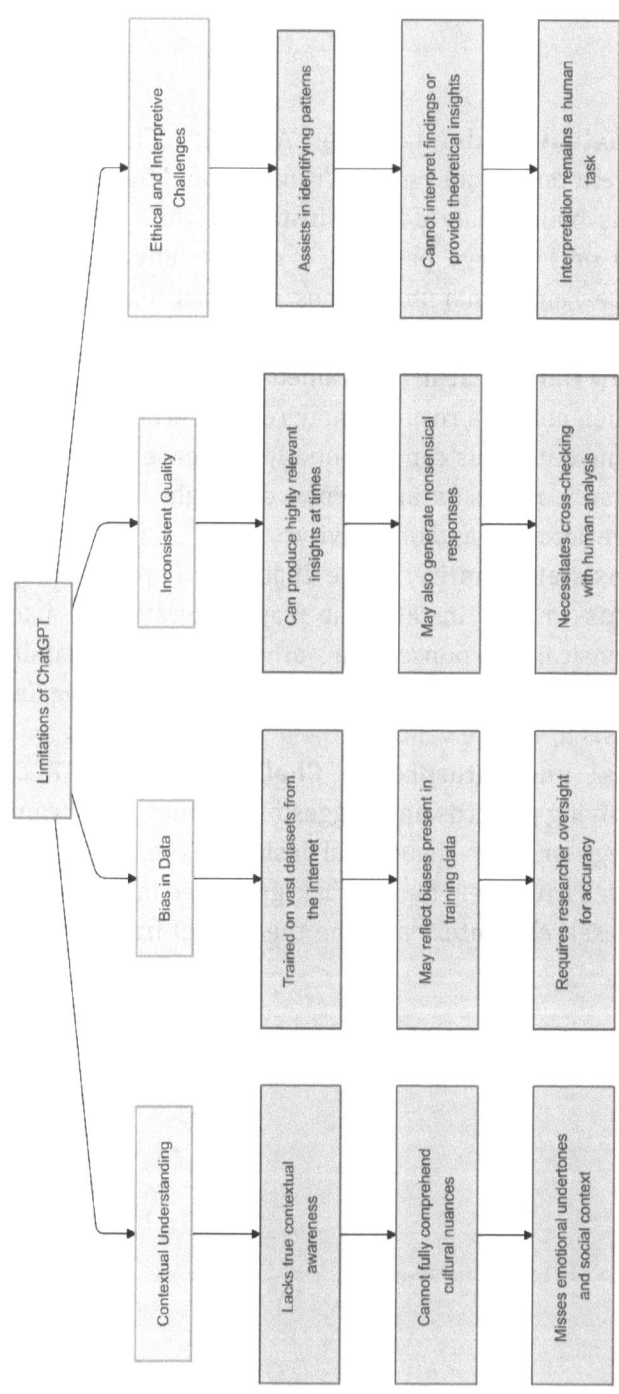

Understanding these limitations helps you maintain realistic expectations for ChatGPT's role in your research and ensures that you use it as a tool to *complement*—not replace—your qualitative analysis process.

Ethical Considerations and Responsible Use

Incorporating ChatGPT into your research introduces several ethical considerations. As with any research tool, it is important to approach AI with responsibility and mindfulness, ensuring that its use aligns with ethical research practices. Here are some key areas to keep in mind:

1. **Data Privacy**: When using ChatGPT, particularly in cloud-based applications, it is essential to ensure that sensitive or personal data remains secure. This is especially crucial in qualitative research, where interview transcripts, focus group notes, and observational data often contain private or confidential information. You should take steps to anonymize the data before using AI tools and ensure that any third-party platforms used comply with data protection regulations, such as GDPR.
2. **Informed Consent**: Participants should be informed if AI tools, like ChatGPT, will be used in the analysis of their data. While AI might streamline the analysis process, participants have the right to know how their information will be processed and who will have access to it. This transparency ensures that ethical standards are upheld and that participants' rights are respected.
3. **Bias and Fairness**: As mentioned earlier, AI tools like ChatGPT can reflect biases present in the training data. In qualitative research, this poses the risk of reinforcing stereotypes or overlooking important but less frequent patterns in the data. Researchers must carefully evaluate AI-generated suggestions and verify them against their own interpretations to ensure fairness and accuracy.
4. **Transparency in AI Contributions**: When presenting findings, it is important to be transparent about the role ChatGPT played in your research. This means clearly documenting how the AI was used— whether it was for coding, summarizing data, or generating themes. Transparency allows your audience to understand the methodology fully and assess the role that AI played in shaping the analysis.
5. **The Human Element**: Finally, it is essential to emphasize that ChatGPT is a *tool*—not a substitute for human judgment. The final

analysis, interpretation of themes, and development of conclusions should always remain in the hands of the researcher. While AI can support and enhance the research process, the depth and meaning of qualitative research come from the researcher's understanding of the context, participants, and theoretical frameworks.

By understanding the basics of AI, the capabilities and limitations of ChatGPT, and the ethical implications of using AI in research, you can begin to harness this powerful tool effectively. As we move forward in the book, we will delve into practical strategies for setting up ChatGPT for your research, creating effective prompts, and integrating AI into your workflow in a responsible and impactful way.

Chapter Key Points:

- *ChatGPT streamlines coding and thematic analysis in qualitative research, saving time and effort.*
- *It summarizes text to identify core ideas from lengthy transcripts.*
- *Assists in text classification and coding by finding recurring words and concepts.*
- *Generates ideas and questions to refine research inquiries.*
- *Recognizes patterns and trends in qualitative data.*

Limitations of ChatGPT:

- *Lacks true contextual understanding of cultural nuances and emotional undertones.*
- *May reflect biases present in its internet-based training data.*
- *Can produce inconsistent or nonsensical responses at times.*
- *Cannot interpret findings or offer theoretical insights; human analysis remains crucial.*

Ethical Considerations:

- *Ensure data privacy by anonymizing sensitive information and complying with regulations like GDPR.*
- *Obtain informed consent from participants regarding the use of AI in data analysis.*
- *Address bias and fairness by critically evaluating AI-generated suggestions.*
- *Be transparent about how ChatGPT was used in the research process.*
- *Emphasize that ChatGPT complements, not replaces, human judgment in qualitative analysis.*

Chapter 3.

THE ROLE OF AI IN QUALITATIVE RESEARCH

Understanding the role of AI begins with a look at the historical context of qualitative methods, the ways in which AI can augment (but not replace) human insight, and addressing common myths about AI taking over the role of researchers. The goal is to highlight the ways in which AI, when used correctly, can enhance qualitative research without compromising its core values of depth, richness, and human interpretation.

HISTORICAL CONTEXT OF QUALITATIVE METHODS

Qualitative research has a long and rich history, originating in fields such as sociology, anthropology, psychology, and education (7–9). The central goal of qualitative research has always been to explore and understand the complexities of human behavior, experience, and social phenomena. Unlike quantitative research, which focuses on numerical data and statistical analysis, qualitative research prioritizes depth, meaning, and context. Methods such as in-depth interviews, focus groups, participant observation, and case studies have been the primary tools for gathering and analyzing qualitative data (10,11).

Traditionally, the process of coding and analyzing qualitative data has been manual, requiring researchers to sift through pages of transcripts, field notes, and other text-based data to identify patterns, themes, and insights. This is labor-intensive, requiring careful interpretation and a deep understanding of the context in which the data was collected. The researcher's role has always been central in qualitative research, with a focus on subjective interpretation, nuance, and the theoretical frameworks that guide analysis.

As the volume of qualitative data grows, especially in the digital age where interviews, social media content, and online discussions can generate massive amounts of text, researchers are increasingly faced with the challenge of managing large datasets while maintaining the quality of their analysis. This is where AI, particularly language models like ChatGPT, has entered the scene—offering new ways to assist with tasks such as coding, pattern recognition, and thematic analysis.

While AI is a relatively new addition to the qualitative research toolkit, it is important to view it as an extension of historical qualitative methods, rather than a replacement. AI does not alter the core goals of qualitative research; instead, it helps researchers manage larger datasets and streamline time-consuming tasks, allowing them to focus more on interpretation and theory-building.

The Synergy Between Human Insight and Machine Assistance

The introduction of AI, and tools like ChatGPT, into qualitative research is best viewed as a collaborative process between human researchers and machine assistance. The synergy between human insight and AI capabilities enhances the research process without diminishing the role of the researcher. In fact, when used properly, AI frees researchers from time-consuming, repetitive tasks, enabling them to devote more energy to the creative, interpretive, and theoretical aspects of their work.

Here is how this synergy plays out in practice:

- **Managing Complex Data Sets**: One of the primary benefits of using AI in qualitative research is its ability to quickly process large volumes of data. For example, ChatGPT can help analyze interview transcripts, focus group discussions, or social media content by identifying recurring themes, patterns, and concepts. However, it is the researcher who interprets these findings, contextualizing them within the framework of the study and deciding which themes are most relevant or significant.
- **Supporting, Not Replacing, Coding**: While ChatGPT can assist in the initial stages of open coding by suggesting possible codes based on recurring words or phrases, it is ultimately up to the researcher to refine these codes, merge them into broader categories, and

interpret their meaning. The AI provides efficiency, but human insight ensures that the analysis captures the richness and complexity of the data.
- **Pattern Recognition and Theme Development**: AI's strength lies in its ability to detect patterns across large data sets that may not be immediately apparent to the human eye. However, AI cannot replace the researcher's deep understanding of the context, culture, and specificities of the data. AI can suggest connections between concepts, but it is the researcher who determines how these connections relate to the theoretical framework or research question.
- **Iterative Analysis**: Qualitative research is often iterative, with researchers refining their analysis over time as new data or insights emerge. ChatGPT can assist in this iterative process by rapidly generating new themes or revising codes as the data evolves. But the key insights still emerge from the researcher's intuition, experience, and theoretical grounding.

The collaboration between human researchers and AI results in a more efficient workflow, but the researcher remains the primary decision-maker. AI serves as a powerful assistant, helping researchers manage data, but it is always the human insight that drives the depth and meaning of the analysis.

DEBUNKING MYTHS ABOUT AI REPLACING RESEARCHERS

There is a common misconception that AI tools like ChatGPT are poised to replace human researchers, especially in fields like qualitative analysis where the process can be time-consuming.

However, this could not be further from the truth.

The idea that AI will render researchers obsolete overlooks the fundamental nature of qualitative research: it is a deeply interpretive process that requires human judgment, cultural understanding, and emotional intelligence—areas where AI still falls short.

Here are some myths about AI replacing researchers and the reality behind them (Figure 3.1).

- **Myth 1: AI Can Fully Automate Qualitative Analysis**
  Reality: While AI can assist in certain stages of the analysis—such as coding or pattern recognition—it cannot replicate the interpretive and contextual understanding that human researchers bring to the process. AI may be able to recognize that certain words frequently appear together, but it does not understand the cultural or emotional nuances behind those words. AI can support the process, but it cannot offer the deep insights that come from human experience and theoretical knowledge.
- **Myth 2: AI Can Replace the Researcher's Role in Interpretation**
  Reality: AI is good at identifying patterns, but it lacks the ability to interpret those patterns within the broader context of a research study. For example, ChatGPT might identify recurring themes in interview transcripts, but it is the researcher who understands how those themes fit into a larger theoretical framework or social context. Interpretation remains a deeply human process, shaped by the researcher's background, knowledge, and interaction with the data.
- **Myth 3: AI Eliminates the Need for Traditional Qualitative Methods**
  Reality: AI does not replace traditional qualitative methods; rather, it enhances them. Researchers still need to conduct interviews, engage in participant observation, or collect narrative data through

traditional methods. AI can assist in the analysis phase, but the data collection and interpretation stages still rely heavily on human interaction and insight.

- **Myth 4: AI Is Flawless**
  Reality: AI is only as good as the data it has been trained on. ChatGPT, for example, may produce biased or incomplete results if the training data is biased or lacks diversity. Researchers must always critically evaluate the suggestions made by AI tools, checking for accuracy, potential biases, and relevance. The role of the researcher is to ensure that AI's contributions enhance the research rather than distort it.

Figure 3.1. Myths about AI in Qualitative Research

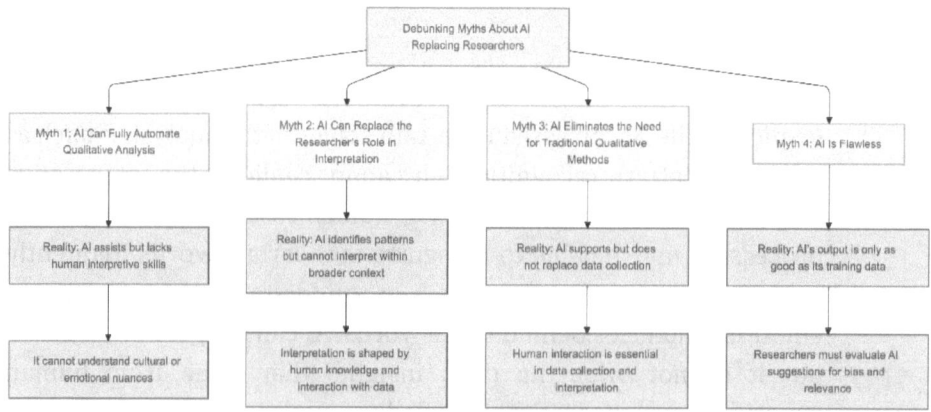

By debunking these myths, we can better understand the true role of AI in qualitative research. Far from replacing researchers, AI tools like ChatGPT are best viewed as assistants—helping manage large volumes of data, suggesting patterns, and streamlining workflows, but always leaving the crucial tasks of interpretation, analysis, and contextual understanding to the human researcher.

AI is an invaluable tool for modern qualitative researchers, but it is most powerful when used in collaboration with human insight. Throughout this book, the goal is to show you how to leverage ChatGPT's strengths while ensuring that you remain in control of the interpretive and analytical processes that make qualitative research unique.

This book is not intended to dive deeply into the various theories and foundations of qualitative analysis.

Instead, the focus is on providing you with a practical guide to how ChatGPT can be utilized to enhance the process of qualitative data analysis. While qualitative research encompasses a wide range of methodologies and theoretical perspectives, my aim here is to demonstrate how AI, specifically ChatGPT, can streamline tasks such as coding, identifying patterns, and aiding in thematic development. By the end of this book, you will understand how to integrate ChatGPT into your qualitative research workflow in a way that complements your existing research skills and methodologies.

For the purposes of demonstration, this book will focus primarily on **thematic analysis** of interview transcripts, and more specifically, **deductive content thematic analysis** based on an already established framework. This type of analysis starts with predefined codes or themes that guide the exploration of the data (12,13), making it particularly suitable for illustrating how ChatGPT can assist in identifying and organizing data according to existing categories. By focusing on this specific method, we will show how ChatGPT can be integrated into your analysis process from the initial coding phase through to the development and refinement of themes.

Chapter Key Points:

- *Qualitative research traditionally involves manual, time-consuming analysis of complex, unstructured data.*
- *Managing large qualitative datasets has become increasingly challenging in the digital age.*
- *AI tools like ChatGPT can assist researchers by streamlining tasks such as coding, pattern recognition, and thematic analysis.*
- *AI serves as an extension of traditional qualitative methods, augmenting but not replacing human insight.*
- *There is a synergy between human researchers and AI, where AI handles data processing and humans provide context and interpretation.*
- *Myths that AI will replace researchers are debunked; AI lacks the cultural understanding and emotional intelligence necessary for deep qualitative analysis.*
- *AI can manage complex data sets and support coding, but final interpretation and theme development remain human tasks.*
- *Ethical considerations emphasize that AI is a tool to support, not replace, human researchers in qualitative analysis.*
- *The chapter focuses on practical guidance for integrating ChatGPT into qualitative research workflows.*
- *By leveraging AI tools like ChatGPT, researchers can enhance efficiency while preserving the depth, richness, and human interpretation that define qualitative research.*

Part II: Practical Strategies for Using ChatGPT

Chapter 4.

INDUCTIVE AND DEDUCTIVE APPROACHES WITH CHATGPT

APPROACHES TO QUALITATIVE DATA ANALYSIS WITH CHATGPT: INDUCTIVE AND DEDUCTIVE APPROACHES

When conducting qualitative data analysis with ChatGPT, researchers can employ two primary approaches: the **inductive approach** and the **deductive approach** (12). Both methods offer distinct advantages and limitations, and their suitability depends on the specific context of the research, the research questions, and the theoretical framework guiding the study. Understanding how each approach works will help you decide which is most appropriate for your research project when using AI assistance. A detailed step by step approach to AI assisted thematic analysis is described in Chapter 10.

INDUCTIVE APPROACH TO QUALITATIVE DATA ANALYSIS

What Is the Inductive Approach?

The **inductive approach** is exploratory and data-driven. In this method, themes, patterns, and codes emerge **from the data itself** rather than being guided by pre-existing theories or frameworks (14,15). Researchers use the inductive approach when they want to discover new insights or understandings directly from the data, without imposing preconceived notions or categories.

When using ChatGPT for inductive qualitative analysis, the AI assists in identifying recurring words, phrases, and patterns across the dataset, which the researcher can then organize into themes. This approach is particularly useful when entering a new or under-researched area where the goal is to develop a theory or understanding based on the data.

Steps for Using the Inductive Approach with ChatGPT:

1. **Input Raw Data**: Feed the raw data (e.g., interview transcripts, focus group notes) into ChatGPT, asking it to identify recurring patterns or concepts.
2. **Generate Initial Codes**: ChatGPT will highlight common words, phrases, or ideas present in the data. These become the initial codes.
3. **Review and Refine**: The researcher reviews ChatGPT's suggestions, selecting which patterns or concepts should be developed into broader themes. This involves significant human judgment to ensure relevance and accuracy.
4. **Theme Development**: From the identified codes, the researcher develops broader themes that explain the underlying meaning of the data. ChatGPT can assist by further elaborating or grouping these themes based on similarities in the text.

Suitability of the Inductive Approach:

- **Exploratory Research**: Ideal for studies that aim to explore a new phenomenon or generate theory rather than test existing hypotheses.
- **Grounded Theory**: Often used in grounded theory approaches, where the researcher builds a theory from the ground up based on emergent data.
- **When Little Pre-existing Knowledge Exists**: Useful in contexts where little is known about the research subject, and the aim is to allow the data to "speak for itself."

Advantages of the Inductive Approach:

- **Flexibility**: No predefined structure allows for the discovery of unexpected insights or themes.
- **Richness of Data**: Since the analysis is open-ended, it captures the full complexity of the data without being constrained by prior assumptions.

- **AI-Assisted Pattern Recognition**: ChatGPT is effective at spotting patterns that may not be immediately obvious to the researcher, aiding in code generation.

Limitations of the Inductive Approach:

- **Over-reliance on AI**: Without predefined categories, there is a risk of over-relying on ChatGPT's suggestions, which might lead to identifying superficial or irrelevant themes.
- **Lack of Focus**: The inductive approach can be time-consuming and may result in an overwhelming amount of data that requires significant refinement by the researcher.
- **Requires Significant Interpretation**: The researcher must still invest considerable time in refining and interpreting the themes, as ChatGPT's outputs may be general or lack the necessary depth for complex analysis.

DEDUCTIVE APPROACH TO QUALITATIVE DATA ANALYSIS

What Is the Deductive Approach?

The **deductive approach** is theory-driven and structured. In this method, the researcher begins with pre-existing themes or categories based on a theoretical framework or prior research, and these are used to guide the analysis (16). This approach is often called **thematic deductive analysis**, where data is coded based on a set of predefined categories or hypotheses.

With ChatGPT, the deductive approach involves asking the AI to identify and categorize the data according to the predefined themes. Rather than discovering new patterns, ChatGPT assists in systematically organizing and sorting the data into the established codes or themes.

Steps for Using the Deductive Approach with ChatGPT:

1. **Define Pre-existing Themes**: Before inputting the data, the researcher defines the themes or categories based on a theoretical model or the research questions.
2. **Tag Data by Themes**: Input the data into ChatGPT and prompt it to categorize responses based on the predefined themes. For example, instruct ChatGPT to tag sections of the text that relate to specific categories like "work-life balance," "team dynamics," or "emotional well-being."
3. **Verify and Refine**: The researcher reviews the AI's output to ensure that the coding is accurate and aligned with the theoretical framework. Adjustments may be necessary to fine-tune the analysis.
4. **Interpretation**: Once the data is sorted, the researcher interprets the findings within the context of the predefined framework.

Suitability of the Deductive Approach:

- **Hypothesis Testing**: Ideal for studies where the goal is to test a specific hypothesis or theory.
- **Structured Research**: Suitable for research projects that are grounded in existing literature and theories, such as when testing a conceptual model or exploring known phenomena.

- **Applied Research**: Often used in applied fields (e.g., psychology, sociology) where specific frameworks or tools (e.g., organizational models, stress frameworks) are used to guide the analysis.

Advantages of the Deductive Approach:

- **Efficiency**: With clear themes already defined, the analysis is more focused and efficient, especially with ChatGPT categorizing data quickly.
- **Alignment with Theory**: The predefined categories ensure that the analysis remains closely aligned with the theoretical framework guiding the study.
- **Consistency**: The use of established themes leads to more consistent and reliable coding across different data sets, especially in large-scale studies.

Limitations of the Deductive Approach:

- **Risk of Missing New Insights**: By focusing on predefined categories, there is a risk of missing emerging themes that were not anticipated by the researcher.
- **Rigid Structure**: The deductive approach can sometimes be too rigid, preventing the researcher from discovering unexpected but valuable insights in the data.
- **Requires Deep Understanding of the Framework**: The researcher must have a strong understanding of the theoretical framework and ensure that it is relevant to the specific data set. ChatGPT's role is limited to organizing the data within those parameters.

INDUCTIVE VS. DEDUCTIVE: WHICH APPROACH TO USE?

When to Use the Inductive Approach:

- **Exploratory Research**: Use the inductive approach when exploring new or uncharted territory where little prior research exists. This allows for flexibility and the potential discovery of new themes or patterns.

- **Theory Development**: Ideal for generating new theories or models, especially in qualitative research areas like grounded theory.
- **Open-Ended Questions**: If your research questions are broad or exploratory, such as "How do employees perceive the impact of virtual reality on team dynamics?" the inductive approach allows the data to guide the findings.

When to Use the Deductive Approach:

- **Hypothesis Testing**: Use the deductive approach when testing a hypothesis or validating a theoretical model. If you're examining something specific—like testing the effect of leadership styles on employee well-being—the deductive approach helps you focus on predefined variables.
- **Applying Existing Theory**: If your study is based on an established theoretical framework, such as Maslow's hierarchy of needs or a stress management model, the deductive approach ensures your analysis aligns with that framework.
- **Focused Research**: For studies that require clear, predefined categories and structured data, the deductive approach keeps the analysis organized and aligned with the research objectives.

LIMITATIONS AND CONSIDERATIONS FOR USING CHATGPT

While ChatGPT can assist in both inductive and deductive approaches, it is essential to understand its limitations:

- **Contextual Understanding**: ChatGPT excels at recognizing patterns in text but lacks true understanding of the context or the nuanced meaning behind the data. The researcher must always review and refine the AI's outputs to ensure they align with the research objectives.
- **Bias in AI Outputs**: ChatGPT may generate biased results based on its training data, and this bias may influence both inductive and deductive analyses. In the inductive approach, AI-generated codes may reflect patterns that are over-represented in the data, while in the deductive approach, ChatGPT's interpretation of predefined themes may not fully capture the richness of the data.

- **Researcher Control**: Regardless of the approach, the human researcher remains essential in interpreting, refining, and applying the findings. ChatGPT assists with the heavy lifting, but final decisions about codes and themes rest with the researcher.

CHOOSING THE RIGHT APPROACH

The decision to use an **inductive** or **deductive** approach to qualitative data analysis with ChatGPT depends on the research goals, the structure of the study, and the existing knowledge of the subject. The **inductive approach** is best suited for exploratory research where new patterns are likely to emerge from the data, while the **deductive approach** is ideal for studies that apply established theories or frameworks to test specific hypotheses. Both approaches, when combined with AI-assisted tools like ChatGPT, can significantly streamline the process of coding and analyzing qualitative data, while the researcher maintains the central role of interpretation and theoretical alignment.

Chapter Key Points:

- *Researchers can use two primary approaches with ChatGPT for qualitative data analysis: inductive (exploratory and data-driven) and deductive (theory-driven and structured).*
- *The inductive approach allows themes and patterns to emerge from the data without predefined theories, making it suitable for discovering new insights in exploratory research.*
- *The deductive approach uses predefined themes or categories based on existing theories or frameworks, ideal for testing hypotheses and applying established concepts.*
- *ChatGPT assists in both approaches by identifying patterns and categorizing data, but lacks true contextual understanding; researchers must review and refine AI outputs critically.*
- *Advantages of the inductive approach include flexibility and capturing the richness of data, while limitations involve potential over-reliance on AI and lack of focus.*
- *Advantages of the deductive approach are efficiency and alignment with theoretical frameworks, but it may miss new insights and can be too rigid.*
- *Choosing between inductive and deductive approaches depends on research goals, existing knowledge, and whether the study aims to explore new phenomena or test specific hypotheses.*
- *Human researchers remain essential for interpretation and final decision-making; ChatGPT serves as an assistant to enhance, not replace, human insight in qualitative analysis.*

Table 4.1 summarizes the key aspects of both inductive and deductive approaches in qualitative analysis with ChatGPT, offering a practical guide for researchers to decide which method best suits their project.

Table 4.1. Key Aspects of Deductive and Inductive Qualitative Analysis

Aspect	Inductive Approach	Deductive Approach
Definition	Data-driven, exploratory approach where themes emerge from the data itself.	Theory-driven, structured approach guided by predefined themes or categories.
Steps for Using ChatGPT	1. Input raw data into ChatGPT. 2. ChatGPT identifies recurring patterns (codes). 3. Researcher reviews and refines codes. 4. Themes are developed from patterns.	1. Define pre-existing themes. 2. Input data and prompt ChatGPT to categorize responses based on predefined themes. 3. Researcher verifies and refines categorizations. 4. Interpret data within the predefined framework.
Suitability	- Exploratory research. - Developing new theories or insights. - Grounded theory or when little prior knowledge exists.	- Hypothesis testing. - Research guided by a specific theoretical framework. - Applied research where categories are already established.
Advantages	- Flexibility to discover unexpected insights. - Allows data to "speak for itself." - ChatGPT helps spot patterns that may be missed by the researcher.	- Efficient and focused analysis. - Ensures alignment with existing theories. - Consistent coding across large datasets.
Limitations	- Can be time-consuming and result in a large amount of data. - Requires significant interpretation and refinement by the researcher. - Risk of over-relying on ChatGPT's pattern recognition.	- May miss emerging themes or novel insights. - Can be too rigid, limiting discovery of new information. - Relies on a strong understanding of the theoretical framework.
When to Use	- When exploring new areas or generating theories. - Research questions are broad or open-ended.	- When testing specific hypotheses or validating a theory. - Research is focused on a structured model (e.g., Maslow's

	- Little prior research exists on the topic.	hierarchy). - Applied research with clear, predefined categories.
Examples of Prompts	*Prompt:* "Analyze this transcript and identify recurring patterns related to team dynamics."	*Prompt:* "Tag all statements related to 'work-life balance' and 'team collaboration' based on these predefined themes."
Researcher's Role	- Significant interpretation is required. - Researcher refines and organizes themes from AI-suggested patterns.	- Researcher reviews and adjusts AI categorizations to fit the theoretical framework.
Best for	- Exploratory studies. - Grounded theory. - When flexibility in analysis is needed.	- Theory-driven research. - Testing hypotheses with structured categories. - Large-scale studies where consistency is important.

Chapter 5.

EFFECTIVE PROMPT ENGINEERING

When conducting qualitative data analysis using ChatGPT, crafting precise, well-structured prompts is critical to obtaining meaningful and relevant outputs. Effective prompt engineering is the foundation of successful AI-assisted analysis, as it guides ChatGPT's responses and ensures that the results are aligned with the research objectives (Table 5.1).

CRAFTING PROMPTS FOR OPTIMAL RESPONSES

The Importance of Well-Crafted Prompts

In qualitative data analysis, the quality of the results you get from ChatGPT largely depends on how well you structure your prompts. A well-designed prompt sets clear expectations and provides the AI with sufficient context to generate a useful response. Since ChatGPT processes information based on patterns and context, vague or ambiguous prompts can lead to irrelevant, superficial, or overly general responses, which can undermine the validity of your analysis.

When crafting prompts, think of it as giving ChatGPT instructions: the clearer and more specific the instructions, the more accurate and focused the AI's output will be.

The following sections outline key principles of crafting effective prompts for ChayGPT (Figure 5.1).

Key Principles for Crafting Effective Prompts

1. **Be Clear and Specific**:
   - Use simple, direct language. Avoid complex or convoluted phrasing that could confuse the AI. For instance, if you are analyzing interview transcripts on the impact of virtual reality (VR) on team collaboration, instead of asking ChatGPT, "What themes are present in this text?", a more specific prompt like, "Identify key themes related to team dynamics and communication in this interview transcript about VR use" will yield more focused results.
2. **Provide Context**:
   - Giving ChatGPT the appropriate context helps it generate responses that are relevant to the research goals. For example, if you are working on a deductive thematic analysis, tell ChatGPT exactly what you need: "I am analyzing the use of VR in remote work environments. Identify any statements related to productivity, emotional well-being, or social interaction."
3. **Break Down Complex Requests**:
   - Avoid overwhelming the AI with overly broad requests. Instead, break the analysis into smaller, manageable tasks. Rather than asking, "Analyze this interview for all key themes," you might ask: "First, identify statements related to team communication. Then, identify statements related to emotional well-being." This structured approach improves the precision of the analysis.
4. **Set Boundaries for Responses**:
   - Sometimes, ChatGPT may return more information than you need or stray from the focus of your analysis. You can limit the scope of its response by including specific instructions. For instance, if you only want ChatGPT to look at a particular theme, specify, "Only focus on themes related to productivity," or "Exclude any comments unrelated to team collaboration."

Example of a Well-Crafted Prompt:

- **Prompt**: "Analyze the following interview transcript and identify themes related to how virtual reality (VR) affects team communication in remote work. Focus specifically on non-verbal cues, social interaction, and team dynamics."
- **Why This Works**: This prompt is clear, provides specific themes of interest, and focuses ChatGPT's attention on the relevant aspects of the transcript, minimizing irrelevant outputs.

Figure 5.1. Key Principles for Crafting Effective Prompts

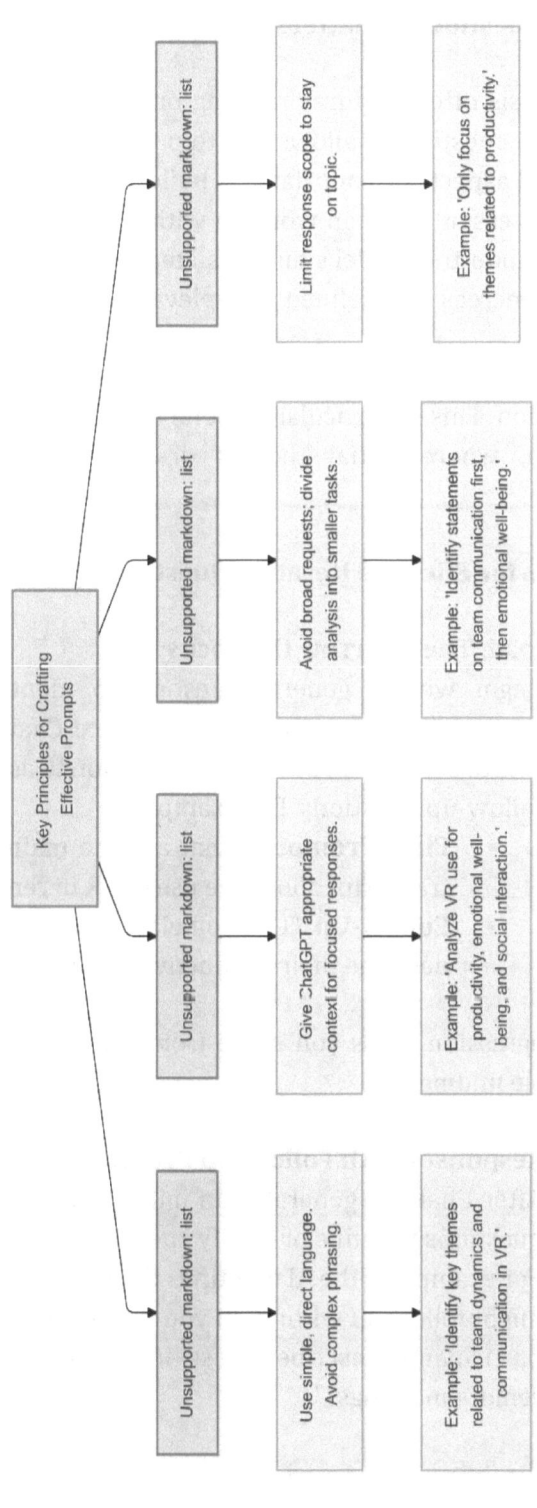

TECHNIQUES FOR ITERATIVE QUESTIONING

Why Iterative Questioning Matters

Qualitative analysis often requires refining or drilling down into specific insights. Iterative questioning allows you to progressively focus on deeper or more nuanced aspects of your data by building on previous responses. This technique is essential when working with AI like ChatGPT because it enables you to guide the model's analysis step by step, ensuring that the outputs become increasingly refined and relevant with each interaction.

With iterative questioning, you treat each round of responses as a basis for further exploration. This is particularly useful for complex or multi-faceted qualitative data, where initial themes or codes may need further elaboration.

Key Techniques for Effective Iterative Questioning

1. **Start Broad, Then Narrow the Focus**:
   - Begin with a general question to identify overarching themes or patterns in the data. Once ChatGPT provides a response, you can use that information to ask more targeted follow-up questions. For example:
     - **First Prompt**: "What are the main themes in this transcript about the use of VR in remote work?"
     - **Follow-Up**: "Can you elaborate on how participants describe their productivity while using VR?"

   This progression helps you move from broad insights to specific, actionable findings.

2. **Refine Responses with Follow-Up Prompts**:
   - After ChatGPT generates an initial response, use follow-up questions to refine or clarify specific aspects of the analysis. For example, if the AI identifies themes related to stress, you might follow up with: "Can you provide more detail on how participants describe stress in relation to using VR for remote meetings?"

This process allows you to delve deeper into specific themes and ensures that the analysis captures the complexity of the data.

3. **Test Multiple Angles**:
   - Qualitative data often contains overlapping themes or nuanced differences in participants' responses. Use iterative questioning to explore these different angles. For example, after identifying general themes related to team dynamics, you could ask:
     - "How do participants describe their relationships with colleagues before and after using VR?"
     - "Do any participants mention changes in how they communicate outside of VR sessions?"

These questions will help uncover patterns or contradictions that might not be evident from the initial response.

4. **Revisit and Clarify Ambiguous Results**:
   - If ChatGPT's initial response is unclear or too broad, use follow-up questions to ask for clarification. For instance, if the AI mentions a theme like "discomfort with technology," you can ask:
     - "What specific aspects of VR technology do participants find uncomfortable?"

This technique ensures that vague results are fleshed out and grounded in the data.

Example of Iterative Questioning:

- **First Prompt**: "Identify the key themes in this transcript regarding participants' emotional responses to using VR in team meetings."
- **Follow-Up Prompt**: "You mentioned stress as a theme. Can you provide specific examples of how participants describe feeling stressed in relation to VR?"
- **Why This Works**: The first prompt identifies broad emotional responses, while the follow-up prompt digs deeper into specific aspects of stress, offering more detailed insights.

HANDLING AMBIGUOUS OR COMPLEX DATA INPUTS

Challenges with Ambiguous Data

Qualitative data is often messy and complex, containing ambiguous or multi-layered responses from participants. Ambiguity may arise from unclear phrasing, contradictory statements, or the use of jargon or slang. In these cases, ChatGPT may struggle to accurately interpret the data unless guided by carefully crafted prompts.

Handling ambiguous data inputs requires a proactive approach. Researchers must anticipate areas of uncertainty and provide ChatGPT with instructions on how to handle conflicting or unclear information.

Strategies for Handling Ambiguous or Complex Data

1. **Clarify Expectations in the Prompt**:
   - When dealing with data that may be ambiguous, specify in the prompt how you want ChatGPT to handle uncertainty. For instance, if participants express conflicting opinions about VR's impact on productivity, you can instruct ChatGPT to identify both sides:
     - "Identify any conflicting viewpoints in this interview about how VR impacts productivity. Highlight both positive and negative statements."
2. **Ask for Multiple Interpretations**:
   - Ambiguous data often requires interpretation. You can prompt ChatGPT to provide multiple possible interpretations to capture the complexity of the data. For example:
     - "The following response from a participant is unclear. Can you provide two possible interpretations of what the participant might mean when they say 'VR is both a barrier and a bridge to communication'?"
3. **Use Clarification Prompts for Complex Data**:
   - If ChatGPT's initial response is too simplistic for complex data inputs, follow up with clarification prompts. For

instance, if a participant mentions mixed feelings about VR, you can ask:
- "Can you elaborate on what the participant means by 'mixed feelings'? Break down the emotional components of their response."

4. **Flag Contradictions**:
   - In some cases, participants may offer contradictory statements. You can prompt ChatGPT to recognize and flag these contradictions for further exploration. For example:
     - "Are there any contradictions in how this participant describes their experience using VR for meetings? If so, summarize them."

Example of Handling Ambiguous Data:

- **Prompt**: "The participant says, 'VR is both empowering and isolating.' What might the participant mean by this? Provide two possible interpretations and explain the nuances of each."
- **Why This Works**: This prompt acknowledges the ambiguity and instructs ChatGPT to explore multiple angles, ensuring a more comprehensive understanding of the participant's response.

Chapter Key Points:

- *Precise, well-structured prompts are essential for obtaining meaningful outputs from ChatGPT in qualitative analysis.*
- *Effective prompts are clear, specific, provide context, break down complex requests, and set boundaries to guide ChatGPT's responses.*
- *Good prompt engineering ensures ChatGPT's outputs align with research objectives and avoids irrelevant or superficial results.*
- *Iterative questioning refines analysis by progressively focusing on deeper aspects of the data.*
- *Techniques include starting broad then narrowing focus, refining with follow-up prompts, testing multiple angles, and clarifying ambiguous results.*
- *Handling ambiguous or complex data requires strategies like clarifying expectations, asking for multiple interpretations, and flagging contradictions.*
- *Researchers must proactively guide ChatGPT to ensure accurate and comprehensive analysis of complex data.*
- *Effective prompt engineering is foundational for successful AI-assisted qualitative analysis, enhancing the precision of ChatGPT's contributions.*

Table 5.1. Effective Prompts Generation for ChatGPT Qualitative Analysis

Section	Key focus	Summary of techniques	Example prompts
1. Crafting prompts for optimal responses	Crafting clear and structured prompts for precise results	1. **Be Clear & Specific**: Use simple language and focus on particular themes. 2. **Provide Context**: Give ChatGPT background info related to research. 3. **Break Down Complex Requests**: Split tasks into smaller queries for detailed analysis. 4. **Set Boundaries**: Limit scope to avoid irrelevant outputs.	*Prompt:* "Identify themes related to how VR affects team communication in remote work, focusing on non-verbal cues."
2. Techniques for iterative questioning	Using a step-by-step approach to refine AI responses	1. **Start Broad, Then Narrow Focus**: Begin with general questions, then get specific. 2. **Refine Responses with Follow-Ups**: Ask follow-up questions for clarification. 3. **Test Multiple Angles**: Explore different viewpoints or facets of a theme. 4. **Revisit Ambiguous Results**: Clarify vague responses.	*First Prompt:* "What are the main themes in this transcript?" *Follow-Up:* "Can you elaborate on how participants describe productivity?"
3. Handling ambiguous or complex data inputs	Guiding AI to handle unclear or contradictory data	1. **Clarify Expectations**: Direct ChatGPT on how to manage conflicting or unclear information. 2. **Ask for Multiple Interpretations**: Prompt AI to provide varied interpretations of ambiguous statements. 3. **Use Clarification Prompts**: Ask ChatGPT to clarify complex data inputs. 4. **Flag Contradictions**: Instruct AI to recognize and flag conflicting statements for further analysis.	*Prompt:* "The participant says, 'VR is both empowering and isolating.' Provide two possible interpretations of this statement."

Chapter 6.

AI-Assisted Open Coding

AI-Assisted Open Coding

Open coding is the foundational step in qualitative data analysis, where raw data is broken down into meaningful segments and assigned codes that represent themes, concepts, or patterns. Traditionally, researchers manually develop these codes, often through repeated reading and analysis of transcripts. However, AI-assisted tools like ChatGPT can significantly enhance and expedite this process by identifying recurring patterns and generating initial codes automatically. In this chapter, we will explore how to effectively use ChatGPT for **AI-assisted open coding**, how to balance **AI-generated codes with researcher intuition**, and the importance of **documenting the coding process for transparency**.

This section lays the groundwork for the step-by-step guide that will follow in Chapter 10, where we will apply these principles to real-world data using the hypothetical interview transcript discussed in Chapter 9. Let us begin by understanding how ChatGPT can help generate initial codes and how to use these codes meaningfully in qualitative research.

What Is Open Coding?

Open coding is the process of breaking down qualitative data—such as interview transcripts, focus group discussions, or observational notes—into smaller segments. Each segment is labeled or "coded" with a term or phrase that summarizes the meaning of that data. In AI-assisted open coding, ChatGPT can help automate the identification of these initial codes, significantly speeding up the coding process.

When you input your data into ChatGPT, the AI scans the text to detect recurring words, ideas, and concepts, which can be used to generate initial codes. These codes serve as the starting point for further analysis, where researchers can refine and group them into broader categories or themes.

How to Use ChatGPT for Initial Code Generation

Before feeding your transcripts into ChatGPT, it is important to ensure that the data is well-prepared by cleaning, formatting, and dividing it into manageable parts, as mentioned in earlier chapters. This allows ChatGPT to function more effectively and accurately detect patterns within the data. Once your data is prepared, you can prompt ChatGPT to assist in the initial coding process by analyzing the transcript and identifying key themes. For instance, if you are examining a transcript discussing the effects of virtual reality (VR) on team collaboration, you could use a prompt like: "Identify key codes in this transcript related to how virtual reality influences team collaboration and communication. Focus on recurring ideas or concepts."

Based on the transcript, ChatGPT will then provide a list of possible codes, which could include terms like "non-verbal cues," "team dynamics," "productivity," or "isolation." While the AI-generated codes can be a great starting point, it is essential to review and refine them to make sure they accurately reflect the content. Since AI-generated codes may be quite broad, refining them through additional questioning can be useful. For example, after receiving the initial codes, you could ask ChatGPT for more specific insights by using prompts like: "You identified 'team dynamics' as a code. Can you break down what the participant said about team dynamics into more specific codes?"

Using ChatGPT to generate preliminary codes can make the initial stages of open coding more efficient, allowing you to spend more time on refining and interpreting the qualitative data.

BALANCING AI SUGGESTIONS WITH RESEARCHER INTUITION

The Role of the Researcher in AI-Assisted Coding

While ChatGPT can provide a significant boost in identifying initial codes, it is important to remember that AI lacks the deep contextual understanding and interpretive skills that human researchers bring to qualitative analysis. Therefore, researchers should treat AI-generated codes as a **starting point** rather than a definitive set of findings.

Balancing AI suggestions with researcher intuition involves reviewing the AI's outputs, assessing their relevance and accuracy, and making decisions about how best to incorporate them into the overall analysis. The AI's role is to assist in identifying patterns, but the researcher remains responsible for ensuring that these patterns are meaningful within the research context.

How to Balance AI Suggestions with Researcher Expertise

1. **Review AI-Generated Codes Critically**:
    - Once ChatGPT generates a list of initial codes, critically assess whether these codes truly represent the core ideas in the data. For example, if ChatGPT identifies "productivity" as a code but does not distinguish between positive and negative aspects of productivity, it is up to the researcher to refine this further by distinguishing between "increased productivity" and "productivity challenges."
2. **Use Researcher Intuition to Refine Codes**:
    - As a researcher, you bring valuable contextual knowledge and understanding to the coding process. Use this expertise to modify or reject AI-generated codes that do not fit your research goals. For example, if ChatGPT generates a code like "technology use," but your study is focused specifically on "virtual reality," you may need to refine or combine these codes to ensure they align with your specific research questions.
3. **Identify Missed Patterns**:
    - AI tools can sometimes miss subtle or nuanced themes that may be critical to your analysis. After reviewing the AI's suggestions, consider whether there are any important

ideas that have not been captured. For example, while ChatGPT may identify broad themes like "stress" or "communication," you might notice underlying emotional nuances like "frustration" or "excitement" that are relevant but were not captured by the AI.

4. **Modify and Merge Codes**:
   o AI-generated codes can sometimes overlap or be too granular. Use your judgment to merge similar codes or group them into broader categories. For instance, ChatGPT might generate codes like "team interaction," "collaboration," and "team dynamics." As a researcher, you could decide to merge these into a single broader category such as "team relationships," or maintain them as separate but related codes depending on your analytical framework.

Balancing AI-generated insights with your own expertise ensures that the final set of codes is both data-driven and theoretically grounded.

DOCUMENTING THE CODING PROCESS FOR TRANSPARENCY

Why Transparency in Coding Matters

In qualitative research, documenting the coding process is crucial for ensuring the rigor and transparency of the analysis. This becomes even more important when using AI-assisted tools like ChatGPT, as reviewers and peers need to understand how the AI was used, what role it played in identifying codes, and how the researcher refined or altered these codes throughout the process.

By documenting each step of the AI-assisted coding process, you create a clear record of how the data was interpreted and coded, which enhances the reliability and credibility of your research. It also allows other researchers to replicate your approach if needed.

Best Practices for Documenting the Coding Process

1. **Describe the Role of ChatGPT in Your Analysis**:
   - Clearly state in your methodology how ChatGPT was used to assist with open coding. For example, explain that ChatGPT was used to generate initial codes, but that all codes were reviewed and refined by the researcher before being applied to the data.
2. **Record Iterative Changes**:
   - If you used iterative questioning to refine the codes, document each step of this process. For instance, if ChatGPT initially generated the code "team dynamics," and you later refined this into "positive team dynamics" and "negative team dynamics," record the rationale behind these changes.
3. **Log Decisions on Code Modification or Rejection**:
   - Keep a log of the codes that you decided to modify, merge, or reject. For each decision, provide a brief explanation. This not only helps in maintaining transparency but also ensures that your coding decisions are well-justified and grounded in the data.
   - Example:
     - **Code Generated by ChatGPT**: "Technology use"

- **Researcher Decision**: Merged into the broader category "virtual reality interactions" to align with the study's focus.
- **Rationale**: The study focuses specifically on VR, and "technology use" was too broad to be useful for this analysis.

4. **Ensure Consistency in Code Application**:
    o Consistency in how codes are applied across different data sets is key to maintaining the integrity of the analysis. Document the rules or criteria you used to apply certain codes so that they are applied consistently across the data. This is especially important if multiple researchers are working on the same project, or if you plan to analyze a large number of transcripts.

5. **Create a Coding Framework or Codebook**:
    o Developing a formal codebook is an excellent way to document the coding process. A codebook contains definitions for each code, examples of how they were applied, and any distinctions made between similar codes. For instance, you could define "team dynamics" as referring to interactions between team members during virtual meetings and provide examples from the transcript to illustrate its application.

SETTING THE STAGE FOR AI-ASSISTED OPEN CODING

AI-assisted open coding with ChatGPT provides a powerful tool for speeding up the initial stages of qualitative analysis. However, it is essential to balance AI-generated codes with researcher intuition to ensure that the final coding scheme is accurate, meaningful, and aligned with the research objectives. Documenting the coding process is critical for transparency and allows for better replication and credibility of the findings.

In the next chapter, I will provide a **step-by-step approach** to applying these principles to the hypothetical transcript from earlier in the book. I will demonstrate how to use ChatGPT to generate initial codes, refine them through iterative questioning, and balance AI suggestions with researcher insight—all while maintaining a clear, transparent documentation process.

Chapter Key Points:

- *Open coding involves breaking down qualitative data into meaningful segments and assigning codes that represent themes or patterns.*
- *ChatGPT can enhance and expedite open coding by automatically identifying recurring patterns and generating initial codes.*
- *Preparing data by cleaning, formatting, and dividing it into manageable parts improves ChatGPT's effectiveness.*
- *Researchers must balance AI-generated codes with their own intuition, critically reviewing and refining the AI's suggestions.*
- *The researcher remains responsible for interpreting the data and ensuring final codes align with research objectives.*
- *Documenting the coding process is essential for transparency and credibility, especially when using AI assistance.*
- *Best practices for documentation include describing ChatGPT's role, recording changes, logging code modifications, ensuring consistency, and creating a codebook.*
- *Combining AI assistance with researcher expertise improves efficiency but requires careful management to maintain rigor in qualitative analysis.*

Chapter 7.

QUALCODER: A FREE, OPEN-SOURCE TOOL FOR QUALITATIVE ANALYSIS

This chapter introduces QualCoder, a free and open-source software designed for qualitative data analysis. By integrating ChatGPT's AI capabilities with QualCoder's robust features, researchers can enhance their analysis workflow, reducing the time spent on manual coding while maintaining analytical depth and rigor.

INTRODUCTION TO QUALCODER

QualCoder is tailored specifically for qualitative research, enabling efficient coding, categorization, and analysis of various data types, including text, images, audio, and video (17). It provides all the essential functionalities required for thorough qualitative analysis, making it an invaluable tool for researchers who desire control over their analysis process without incurring costs (18–21).

Whether you are conducting interviews, analyzing focus group transcripts, or coding visual data, QualCoder helps you systematically organize and explore your data. It supports multiple data formats like DOCX, ODT, PDFs, images, and multimedia files. With features such as hierarchical coding structures, coder comparison, and visual reporting tools, QualCoder offers powerful functionality combined with simplicity, making it suitable for both new and experienced researchers.

OVERVIEW OF QUALCODER'S FEATURES

QualCoder stands out as a feature-rich platform without the expense of commercial software. Its open-source nature ensures accessibility for all users. Key features include:

1. **Text and Multimedia Coding**: Allows coding of text, images, video, and audio files, expanding the scope of analysis across different data types—beneficial for fields that rely on non-textual data.
2. **Hierarchical Code Categorization**: Enables grouping of codes into a tree-like structure, providing an organized way to manage complex coding schemes essential for large datasets or multifaceted themes.
3. **Memoing and Journaling**: Supports creating memos and journal notes to document thoughts, reflections, and interpretations throughout the research process, enhancing transparency and credibility.
4. **Visual Reports and Coder Comparisons**: Generates various reports, including code frequency, coding graphs, and coder comparisons using statistical measures like Cohen's Kappa to assess inter-coder reliability. These tools aid in presenting findings and understanding relationships between codes.
5. **Cross-Platform Compatibility**: Runs on Windows, macOS, and Linux, facilitating collaboration with team members across different operating systems without compatibility issues.
6. **Community-Driven Development**: Continuously improved by its user community, with opportunities to suggest features and report bugs via GitHub, ensuring the software evolves to meet user needs.
7. **Offline Functionality**: Operates without internet connectivity, providing researchers full access to their projects in fieldwork settings or areas with limited internet access.
8. **Export and Import Options**: Supports various file formats for import and export, including text, CSV, and HTML, and aims to support the REFI-QDA standard for interoperability with other qualitative software, promoting interdisciplinary and collaborative work.

Why Choose Open-Source Software for Qualitative Analysis

Open-source tools like QualCoder offer significant advantages over expensive commercial software, especially for researchers and institutions with limited budgets. They provide high-quality qualitative analysis capabilities without financial barriers, making them accessible to students, early-career researchers, and independent scholars.

1. Cost Efficiency

QualCoder is completely free, eliminating the expensive fees or subscriptions required by many qualitative data analysis tools. This affordability democratizes access to essential research tools, allowing more researchers to engage in qualitative inquiry without budget constraints.

2. Customization and Control

As an open-source software, QualCoder allows users to modify and tailor the program to meet specific research needs—a flexibility rarely available in commercial software. Researchers with technical expertise can enhance functionalities or add new features, making it especially valuable for complex, interdisciplinary projects where standard software may not suffice.

3. Transparency and Data Ownership

With QualCoder, researchers maintain full ownership and control over their data without concerns about restrictive licenses or institutional ties. The software operates entirely offline, ensuring data security and eliminating reliance on cloud-based services that might restrict data access.

4. Community Support and Continuous Improvement

QualCoder benefits from a collaborative user community that contributes to its development by suggesting new features, reporting bugs, and enhancing functionalities. This community-driven model ensures the software evolves in response to real-world research needs, keeping it up-to-date with new functionalities and improvements.

Benefits of Integrating QualCoder into Your Research Workflow

QualCoder integrates smoothly into research workflows for both individual researchers and collaborative teams, offering several key benefits:

1. Streamlined Coding Process

Its intuitive interface and flexible coding options simplify the organization and management of complex qualitative data. Whether analyzing text interviews or multimedia content, QualCoder allows researchers to focus on interpretation rather than navigating complicated software interfaces.

2. Collaborative Flexibility

While primarily designed for single-user projects, QualCoder supports collaboration by allowing projects to be shared between researchers. Multiple coders can work independently on the same dataset and later compare their coding using built-in tools, enhancing collaboration without compromising individual coding efforts.

3. Efficient Data Management

QualCoder can handle multiple data types—including text, audio, and video—within a single project. This capability eliminates the need to switch between different software for various media types, allowing researchers to categorize and code all their data in one place.

4. Enhanced Rigor with Visual Reports

The software's reporting features, such as coding frequency charts, code hierarchy graphs, and coder comparison metrics, provide clear and structured ways to present and validate findings. These visual tools enhance the rigor of research by offering transparent documentation of the analysis process.

5. Long-Term Sustainability

As an actively developed open-source tool, QualCoder ensures long-term project sustainability. There's no risk of losing access due to expired licenses or pricing changes, and its compliance with open standards like REFI-QDA allows for seamless data exchange with other qualitative software, making it future-proof.

Getting Started with QualCoder

Before beginning your qualitative analysis with QualCoder, it is important to install and configure the software properly. This section offers a step-by-step guide to help you download, install, create your first project, and navigate the interface. Whether you are an experienced researcher or new to qualitative data analysis, QualCoder's straightforward setup allows you to start working with your data quickly.

Downloading and Installing QualCoder

1. **Access the GitHub Repository**: Visit https://github.com/ccbogel/QualCoder to find the latest version of the software.
2. **Select the Appropriate Version**:
   - **Windows Users**: Download the Windows installer executable (.exe) from the Releases section.
   - **macOS Users**: Look for the macOS executable or follow instructions to build from the source code.
   - **Linux Users**: Download the .tar.gz file or use your package manager to install dependencies and run QualCoder from the source.
3. **Install the Software**:
   - **Windows**: Run the downloaded .exe file and follow the on-screen instructions. If you plan to code audio or video files, install the VLC media player.
   - **macOS and Linux**: You may need to install additional dependencies like Python and VLC. Detailed installation instructions are available on the GitHub Wiki and documentation page.
   - **Python Installation**: For source code installations, ensure Python 3.10 or later is installed, along with the required Python packages listed in the requirements.txt file.
4. **Launch QualCoder**: After installation, open QualCoder from your applications menu. The software should start quickly, ready for you to create your first project.

System Requirements and Compatibility

QualCoder is accessible across various platforms with minimal system requirements:

- **Operating Systems**:
    - **Windows**: Compatible with Windows 10 and later.
    - **macOS**: Works on current macOS versions; manual installation of dependencies may be required.
    - **Linux**: Supports distributions like Ubuntu, Fedora, and Arch Linux.
- **Minimum Requirements**:
    - **Python 3.10 or Later**: Necessary if installing from the source code.
    - **VLC Media Player**: Required for coding audio and video files.
    - **Screen Resolution**: At least 1024 x 600 pixels recommended.
    - **Disk Space**: Ensure sufficient storage for your data, especially multimedia files.
- **Offline Functionality**: QualCoder operates entirely offline, making it ideal for environments with limited internet access or for handling sensitive data securely.

Initial Setup: Creating a Project and Navigating the Interface

1. **Create a New Project**:
    - **Start the Program**: Upon opening QualCoder, choose **Create New Project**.
    - **Name Your Project**: Enter a meaningful project name that reflects your research.
    - **Choose a Location**: Select a folder to save your project files.
    - **Save the Project**: Click **Create** to initialize your project workspace.
2. **Navigate the Interface**:
    - **Menu Bar**: Access primary functions like project management, file import, report export, and preferences.

- o **Project Explorer Panel**: Located on the left, this panel lists your imported files and lets you navigate between data types (text, images, audio, video).
- o **Coding Area**: The central pane where you view and code your data. Select text or regions in media files to assign codes.
- o **Code Tree Panel**: On the right, this displays your codes and their hierarchical relationships, allowing you to manage and organize them.
- o **Memos and Journals**: Additional tabs for creating notes and reflections to document your analysis process.

3. **Set Preferences and Parameters**:
   - o **Language and Appearance**: Adjust settings for interface language, font sizes, and colors in the Preferences menu.
   - o **Saving and Backup**: QualCoder auto-saves your work. Set up manual backups for extra security, and consider regular backups to external storage.

IMPORTING DATA INTO QUALCODER

Before you can begin analyzing your qualitative data in QualCoder, the first step is to import your research materials into the software. QualCoder supports a wide range of file types, including text documents, PDFs, images, audio, and video files, allowing you to work with diverse data sources. This section will guide you through preparing your data for import, walk you through the steps of uploading files, and explain how to organize and manage your data within your project.

Preparing Data for Import (Transcripts, Text Files, and Media)

Properly preparing your data before importing it into QualCoder ensures that the analysis process runs smoothly. Data preparation is critical for maintaining consistency, improving readability, and making sure that the software can correctly interpret different file formats.

1. Cleaning and Formatting Text Data

- **Transcripts**: If you are working with interview or focus group transcripts, ensure that they are clean and free from formatting errors. Remove unnecessary text (e.g., time stamps or interviewer prompts) unless it is essential for analysis. Each speaker should be clearly indicated (e.g., "Interviewer" and "Participant"), and the text should be consistently formatted for easy coding.
- **Text Files**: QualCoder supports various text formats such as .txt, .odt, .docx, .pdf, .html, and .epub. Ensure that text files are complete, readable, and properly formatted before importing them.
- **Standardizing Text Data**: For projects involving multiple files, it's essential to standardize how text is presented across different files. This might include ensuring consistent fonts, margins, and speaker labels, so the text is easy to code.

2. Preparing Multimedia Files (Audio, Video, and Images)

- **Audio and Video Files**: If your research includes multimedia files (e.g., recorded interviews or focus group discussions), ensure the files are in a format supported by QualCoder. The software requires the VLC media player for audio and video playback, so make sure it's

installed on your computer. Commonly supported formats include .mp3, .wav (audio), and .mp4, .avi (video). Trim unnecessary parts of the files (e.g., long silences or irrelevant sections) to make coding more efficient.
- **Images**: QualCoder supports image files such as .jpg, .png, and .bmp. Ensure that images are high quality, properly cropped, and relevant to your research before importing them.

3. File Naming Conventions

- To keep your data organized, use clear and consistent naming conventions for all your files. For example, label interview files by participant name or number (e.g., "Interview_Participant_1.docx") and date of data collection. Consistent naming makes it easier to manage files within QualCoder and track progress during analysis.

Step-by-Step Guide to Importing Text, PDFs, and Multimedia

Once your data is cleaned and ready, the next step is to import it into QualCoder for analysis. The process is user-friendly and allows you to work with multiple file types, making QualCoder a versatile tool for different kinds of qualitative research.

1. Importing Text Files

- **Open Your Project**: Once you have launched QualCoder and created a project, go to the **Files** menu located at the top of the screen.
- **Choose Import Text**: Select **Import Text** from the dropdown menu. This will open a file browser where you can navigate to the location of your text files.
- **Select the Files**: Choose the .txt, .odt, .docx, or other supported text files you wish to import. You can select multiple files at once if you are importing several transcripts or documents.
- **Confirm Import**: Click **Open** to import the selected files into your project. The files will now appear in the **Project Explorer** panel, ready for coding.

2. Importing PDF Files

- **PDF Import**: QualCoder also supports the import of .pdf files, which is useful if your qualitative data comes in this format (e.g., reports, articles, or scanned transcripts). The steps for importing PDFs are similar to text files:
    - Go to the **Files** menu, then select **Import PDF**.
    - Browse your files, select the relevant PDFs, and click **Open**.
    - Once imported, the PDF files will appear in the Project Explorer, and you can begin coding them similarly to text.

3. Importing Multimedia (Audio, Video, and Images)

QualCoder is unique in its ability to handle audio and video files, making it a versatile tool for multimodal qualitative research.

- **Audio/Video Import**: To import audio or video files:
    - Navigate to the **Files** menu and select either **Import Audio** or **Import Video**.
    - Browse for your files (e.g., .mp3, .wav, .mp4), select them, and click **Open**.
    - Once imported, you can code specific time segments in the media player embedded within QualCoder. You can also create memos or notes directly linked to sections of the media files.
- **Image Import**: Importing images is done through the **Import Image** option under the **Files** menu.
    - After importing .jpg, .png, or other supported image files, you can annotate and code specific parts of the images.
    - Like text and audio/video, images will be displayed in the Project Explorer panel for easy access.

4. Confirming the Import

Once the files are imported, you can view all your documents in the Project Explorer panel on the left side of the screen. You should check that the files are fully imported and correctly formatted for coding. You can open each file to ensure the text is readable, the media plays smoothly, and images are correctly displayed.

ORGANIZING AND MANAGING DATA WITHIN PROJECTS

After importing your data, effectively organizing and managing your files is critical for streamlining your analysis workflow, especially when working with large datasets. QualCoder provides various tools to help you manage and structure your data within projects.

1. Project Explorer: Managing Files

- **File Structure**: All imported files are listed in the Project Explorer panel. You can easily navigate between text, PDF, image, audio, and video files from this panel. The Project Explorer acts as a central hub for accessing and managing your data.
- **Folders and Sub-Folders**: Organize your files by creating folders and sub-folders to group similar data. For example, you can create folders for different data types (e.g., "Transcripts," "Audio Interviews") or by participants, themes, or phases of your study.

2. Assigning Codes and Categories

- Once your data is imported, you can start assigning codes to various segments. QualCoder allows you to organize your codes into hierarchical categories, which can be useful when dealing with complex or large datasets. Codes can be nested under broader categories, enabling a structured approach to data analysis.
- For example, you might create a parent category called "Workplace Challenges" with subcategories like "Communication Issues," "Workload Pressure," and "Job Satisfaction." This structure makes it easier to track and analyze the relationships between different themes.

3. Creating Memos and Journals

- **Memos**: Throughout your analysis, you can create memos to document your observations, reflections, and interpretations. Memos can be attached to specific files, codes, or even parts of text or multimedia, helping you keep track of key insights.
- **Journals**: QualCoder also allows you to maintain a research journal, where you can log your thoughts about the project as it progresses.

Journals are useful for documenting decisions, adjustments in methodology, or emerging patterns in the data.

4. Managing Coder Comparisons

- If you are working with a team of coders, QualCoder offers tools to compare coding between different researchers. You can run reports to assess coding agreement, which can be quantified using metrics like Cohen's Kappa to evaluate inter-coder reliability.
- This feature is particularly useful when you need to demonstrate the rigor and consistency of your coding process in collaborative projects.

5. Backing Up and Exporting Data

- **Backup Files**: Regularly back up your project to prevent data loss. QualCoder projects are stored in a folder containing all the coding data, which can be copied to an external drive or cloud storage.
- **Exporting Data**: You can export coded data and reports in various formats, including .csv, .txt, and .html. This is useful for sharing findings with colleagues, preparing reports, or transferring data to other qualitative analysis software.

CODING AND CATEGORIZATION IN QUALCODER

Coding is a fundamental aspect of qualitative data analysis. It involves identifying meaningful segments of data and labeling them with codes that represent themes, patterns, or ideas emerging from the content. QualCoder offers robust tools for both manual and AI-assisted coding, providing flexibility to researchers who want to use traditional coding methods or speed up the process with AI. This section covers how to approach manual and AI-assisted coding in QualCoder, how to create and manage codes and categories, and how to navigate the software's intuitive coding interface.

Manual Coding vs. AI-Assisted Coding: How to Use Both Effectively

QualCoder primarily supports **manual coding**, where researchers review the data and apply codes based on their insights and interpretive framework. However, QualCoder can also be integrated with **AI-assisted coding** tools, such as ChatGPT, to enhance the coding process by suggesting patterns or themes from the data.

1. Manual Coding

Manual coding involves researchers closely reading or reviewing their data and assigning codes to segments of text, images, audio, or video that reflect key ideas or themes. This method allows for deep engagement with the data and ensures that the researcher's understanding and theoretical framework guide the coding process.

- **Advantages**:
    - Provides rich, in-depth insights because researchers immerse themselves in the data.
    - Encourages reflection, memoing, and journaling, which help maintain transparency in the research process.
    - Allows for nuanced interpretation and application of codes that reflect the subtleties of language, emotion, and context.
- **Disadvantages**:
    - Time-consuming, especially for large datasets.
    - Risk of researcher bias if the coding process is not rigorously checked or triangulated.

2. AI-Assisted Coding

While QualCoder does not have built-in AI capabilities for auto-coding, it can be integrated with AI tools like ChatGPT to assist with generating codes or suggesting themes based on the text. AI-assisted coding is particularly useful for identifying initial patterns across large datasets, which can then be refined through manual review.

- **Advantages**:
    - AI can process large volumes of data quickly, identifying initial themes or patterns that may not be immediately obvious to human coders.
    - Reduces the workload for researchers when dealing with repetitive or massive datasets.
    - Provides an opportunity to cross-check the results of manual coding, ensuring comprehensive theme coverage.
- **Disadvantages**:
    - AI may overlook context-specific nuances or cultural subtleties that are crucial in qualitative research.
    - Risk of bias or over-reliance on AI-generated suggestions, which may not align with the research's theoretical framework.
    - Requires researcher oversight to validate AI-generated codes.

3. Using Both Methods Effectively

For best results, combine AI-assisted coding with manual coding:

- Start by using AI to scan the dataset and suggest broad patterns or recurring terms. AI can help you quickly identify potential themes that you may explore further.
- Once the AI provides an initial set of codes or themes, review these manually. Refine or discard AI-generated codes that do not fit your research framework or that lack nuanced understanding.
- Use manual coding to apply deeper, context-specific interpretation to the data, ensuring that your theoretical lens guides the analysis.

- By blending AI's efficiency with manual coding's depth, you can expedite the process while ensuring accuracy and theoretical alignment.

Creating Codes and Categories

In QualCoder, codes and categories form the backbone of your qualitative analysis. Codes are labels you assign to specific segments of your data, while categories group related codes into broader themes. Organizing codes into categories helps you structure your analysis and reveal patterns in the data.

1. Creating Codes

- **Selecting Text or Data Segments**: When you are reviewing a file (whether it is a text, image, audio, or video), you can highlight specific portions of the data to apply a code.
    - For text: Highlight the relevant words, sentences, or paragraphs.
    - For images: Select a portion of the image that reflects the theme you want to code.
    - For audio or video: Select the time segment you want to code.
- **Assigning a Code**: Once the segment is selected, right-click to access the coding options and select **Add New Code**. Enter a descriptive name for the code that reflects the content's theme or meaning.
    - **Descriptive Codes**: These codes capture surface-level content, such as "Job Satisfaction" or "Work-Life Balance."
    - **Interpretive Codes**: These codes reflect deeper, more abstract interpretations of the data, such as "Emotional Resilience" or "Hidden Power Dynamics."
- **Memoing and Reflections**: After assigning a code, you can add memos (notes) to document your reflections on why this code was applied or what stood out in the data. Memos are useful for maintaining transparency and helping you track your evolving thoughts throughout the analysis.

2. Creating Categories

- **Organizing Codes into Categories**: Once you have created several codes, you can group related codes into categories to create a hierarchical structure. For example, if you have individual codes like "Job Stress," "Communication Issues," and "Teamwork Problems," you can group them under a broader category like "Workplace Challenges."
- **Parent and Child Codes**: Categories often take the form of parent (broad category) and child (specific code) relationships. This helps structure your data in a meaningful way, making it easier to identify patterns at both the macro and micro levels of your analysis.
- **Creating New Categories**: To create a category, right-click on the **Code Tree Panel** (usually located on the right side of the screen) and select **Add New Category**. You can then assign existing codes to this category, building out a thematic structure.
- **Dynamic Categories**: As your analysis progresses, you may find that new categories or sub-categories emerge. QualCoder allows you to easily move codes around, refine categories, and continuously develop your code structure as your understanding of the data deepens.

Navigating the Coding Interface: Code Trees, Memos, and Annotations

The coding interface in QualCoder is designed to make it easy to manage your codes, view relationships between themes, and document your thoughts. The primary components you will use are the **Code Tree**, **Memos**, and **Annotations** features.

1. Code Tree Panel

- **Overview**: The Code Tree Panel displays all the codes and categories you have created in a hierarchical tree format. This panel makes it easy to visualize how your codes are related and navigate between them.
- **Managing Codes and Categories**: You can add, rename, or delete codes and categories directly from the Code Tree. To move a code into a different category, simply drag and drop it. This flexibility is

especially useful as new themes emerge and require adjustments to your coding structure.
- **Color Coding**: QualCoder allows you to assign colors to different codes, which can make it easier to visually differentiate themes when reviewing coded text or images. This is particularly helpful for identifying recurring themes at a glance.

2. Memos

- **Linking Memos to Codes**: You can attach memos to individual codes or categories, documenting your reflections on why a certain code was applied or what insights the code reflects. Memos provide an important audit trail that demonstrates the thought process behind your coding decisions.
- **General Memos**: In addition to code-specific memos, you can create general memos to document your overall impressions or emerging ideas as you progress through the analysis. These can serve as research journals, helping you reflect on the bigger picture and the trajectory of your project.

3. Annotations

- **Adding Annotations to Data**: Annotations allow you to leave comments directly on segments of text, images, or multimedia. These annotations act as reminders or prompts for further exploration and can also serve as markers for returning to particularly important parts of the data.
- **Linking Annotations to Codes**: While coding a segment, you can add annotations that explain why this part of the data was coded in a particular way. This not only enhances the transparency of your coding decisions but also helps you stay organized, particularly when working with large datasets.
- **Reviewing Annotations**: All annotations can be accessed and reviewed at any time, providing you with a quick way to revisit important insights or areas that need further analysis.

QualCoder's coding interface, combined with its flexible tools for creating and managing codes and categories, empowers researchers to conduct detailed and systematic qualitative analysis.

ANALYZING DATA IN QUALCODER

Once you have completed the coding and categorization of your qualitative data, the next step is to dive into the analysis. QualCoder provides a range of built-in analysis tools that help you uncover deeper patterns, relationships, and insights from your data.

Overview of Built-In Analysis Tools

QualCoder comes equipped with several powerful tools that allow you to analyze qualitative data from various angles. These tools help you explore relationships between codes, track the frequency of themes, and assess coder agreement in collaborative projects.

1. Word Frequency Analysis

Word frequency analysis is one of the most fundamental tools in QualCoder. This tool counts the occurrence of individual words within your dataset and displays the most frequently used words in a ranked list. This is particularly useful for identifying dominant themes or terms that might warrant further investigation.

- **How to Use**: Navigate to the **Analysis** menu and select **Word Frequency**. Choose which data sources (e.g., individual transcripts or the entire dataset) you want to analyze. The results will display a list of words with their frequency counts.
- **Applications**: Use word frequency analysis to get an initial sense of the prominent language patterns in your data. This can help you identify common concerns, emotions, or recurring themes across interviews, focus groups, or documents. You can also exclude common words (like "the," "and," "but") using stop lists to focus on meaningful terms.

2. Code Co-occurrence Tables

Co-occurrence tables allow you to examine the relationships between different codes in your dataset. This is a critical tool for understanding how often two or more themes appear together in the same segment of data, which can provide insights into complex interconnections between ideas or issues.

- **How to Use**: Go to the **Analysis** menu and choose **Co-occurrence Tables**. Select the specific codes or categories you want to compare, and QualCoder will generate a table showing how frequently those codes overlap in your data.
- **Applications**: Co-occurrence analysis is particularly useful when exploring how themes interact. For example, if you are studying workplace dynamics, you might want to see how often "Stress" co-occurs with "Leadership Issues" or "Work-Life Balance." This can reveal deeper, more integrated patterns that may not be apparent through individual coding alone.

3. Coding Frequencies

Coding frequency reports provide a count of how many times each code or category has been applied across your dataset. This tool gives you an overview of the distribution of your codes, helping you identify which themes are more prevalent and which are less frequent.

- **How to Use**: Select **Coding Frequencies** from the **Reports** menu. You can choose to display results for individual codes or broader categories, and you can filter by file or across the entire dataset.
- **Applications**: Use coding frequencies to assess the prominence of different themes in your research. If you are analyzing focus groups, for example, this report will show which topics were most frequently discussed across participants.

4. Coder Comparison

If you are working with multiple coders, the **Coder Comparison** tool helps evaluate inter-coder reliability by comparing how different researchers applied codes to the same dataset. The tool uses the **Cohen's Kappa** statistic to measure agreement between coders, providing a quantitative measure of reliability in qualitative research.

- **How to Use**: Navigate to the **Analysis** menu and select **Coder Comparison**. Choose the data segments coded by different researchers, and the tool will calculate the level of agreement.
- **Applications**: Coder comparison is invaluable in collaborative projects where multiple coders analyze the same data. High agreement suggests that the coding process is consistent and reliable, whereas discrepancies may indicate areas where further discussion or clarification is needed.

5. Text Search

The text search tool allows you to locate specific words or phrases across your dataset, helping you quickly find relevant data segments for further analysis or coding. This is particularly useful when you want to track how a specific term or topic is used throughout your data.

- **How to Use**: Go to the **Analysis** menu and choose **Text Search**. Input the word or phrase you are looking for, and QualCoder will highlight all instances where it appears in the data.
- **Applications**: This tool is helpful for thematic tracking, especially if you are looking to explore how a key concept (e.g., "innovation" or "resilience") is discussed across multiple interviews or documents.

Visualizing Data: Word Clouds, Code Relationships, and Graphs

Visualization is a powerful way to explore and communicate your findings, particularly when presenting complex qualitative data to audiences who may not be familiar with the nuances of your research. QualCoder includes several visualization tools that help you understand and present your data in a clear and engaging way.

1. Word Clouds

Word clouds provide a visual representation of word frequency data. The more frequently a word appears in your dataset, the larger it appears in the word cloud. This tool offers a quick, visually engaging way to identify key themes in your data.

- **How to Use**: After running a word frequency analysis, you can generate a word cloud directly from the results. QualCoder will create a cloud where the most prominent words are displayed in a larger font size, providing an immediate sense of which terms dominate your dataset.
- **Applications**: Word clouds are particularly useful for presentations and reports, offering an intuitive way for stakeholders to grasp the major themes without needing to sift through detailed reports.

2. Code Relationship Graphs

Code relationship graphs allow you to visualize how different codes are connected based on co-occurrence. These graphs display codes as nodes and relationships as edges, showing how frequently certain themes appear together in your data.

- **How to Use**: Navigate to the **Visualizations** menu and select **Code Relationship Graph**. Choose the codes you wish to include, and QualCoder will generate a graph that shows how the selected codes are interconnected.
- **Applications**: Use code relationship graphs to visualize complex interactions between themes, such as how "Workplace Stress" connects with "Job Satisfaction" and "Management Support." This visual tool can help you see underlying structures in your data that may not be apparent through text-based analysis alone.

3. Hierarchical Code Trees

Hierarchical code trees visualize the structure of your codes and categories. If you have organized your codes into parent and child relationships, the code tree will display these connections, helping you understand how specific sub-themes relate to broader categories.

- **How to Use**: In the **Visualizations** menu, select **Hierarchical Code Tree**. QualCoder will generate a tree diagram showing the parent-child relationships between your codes.
- **Applications**: Code trees are helpful for presentations where you need to explain the thematic structure of your research. For example, a tree might show how "Workplace Challenges" branches

into "Communication Issues" and "Teamwork Conflicts," giving a clear visual representation of your analytical framework.

Exporting Data and Results for Further Analysis

After completing your analysis in QualCoder, you may need to export your data and results for further analysis, collaboration, or reporting. QualCoder offers several export options to ensure that your findings can be easily shared or integrated with other tools.

1. Exporting Coded Data

- **How to Export**: Go to the **Files** menu and select **Export**. You can choose to export your entire coded dataset or specific portions, such as individual transcripts or files. QualCoder supports export formats like .csv, .txt, .odt, and .html.
- **Applications**: Exporting your coded data is essential for sharing results with collaborators, submitting findings for peer review, or transferring data to other qualitative analysis software for further analysis. You might also export to .csv if you wish to analyze the data in a statistical program such as SPSS or R.

2. Exporting Reports

- **Coding Frequency Reports**: You can export coding frequency reports to display how often specific codes were applied across the dataset. These reports are available in .csv and .txt formats for easy integration into spreadsheets or other reporting tools.
- **Word Frequency Reports**: Similarly, word frequency reports can be exported in .csv format, making it easy to incorporate them into detailed analysis or share them with colleagues.
- **Code Co-occurrence Reports**: Co-occurrence tables can also be exported to .csv for further exploration or inclusion in your research outputs.

3. Exporting Visualizations

- **How to Export**: Visualizations, including word clouds, code relationship graphs, and code trees, can be exported as image files (e.g., .png) for easy inclusion in reports, presentations, or publications.

Applications: Exported visualizations are useful for presenting your findings to stakeholders or illustrating key themes in academic papers. They offer a visually compelling way to communicate complex qualitative data.

Chapter Key Points:

- *QualCoder is a free, open-source qualitative analysis software with robust features comparable to commercial tools, offering cost efficiency and allowing customization to meet specific research needs.*
- *It supports coding and analysis of various data types—including text, images, audio, and video—with features like hierarchical code categorization, memoing, journaling, visual reports, and coder comparison tools, enhancing versatility and depth in analysis.*
- *QualCoder is cross-platform compatible, running on Windows, macOS, and Linux, which facilitates collaboration among researchers using different operating systems.*
- *The software ensures data ownership and operates entirely offline, giving researchers full control over their data without concerns about restrictive licenses or cloud-based storage.*
- *Integrating QualCoder into research workflows streamlines the coding process, enhances collaboration, improves data management, and provides long-term sustainability due to its active development and compliance with open standards.*
- *Getting started with QualCoder involves downloading and installing the software from its GitHub repository, creating a new project, and navigating its user-friendly interface to begin importing and analyzing qualitative data.*

Chapter 8.

INTEGRATING CHATGPT WITH QUALCODER

This chapter explores how to combine ChatGPT's AI capabilities with QualCoder, a powerful, free, open-source tool for qualitative data analysis. The integration of these two platforms can significantly streamline the research process, reducing the time spent on manual coding while ensuring that the depth and rigor of high-quality research is maintained.

There are three main approaches to integrating ChatGPT-generated codes into QualCoder, each offering varying levels of automation depending on the researcher's technical expertise and project requirements.

Note: Scope of This Book

This book focuses on the Manual Import approach for integrating ChatGPT with QualCoder as outlined in detail in Chapter 10. Although semi-automated and fully automated approaches offer increased automation, they are outside the scope of this guide. For practical and operational purposes, the manual approach is emphasized due to its accessibility and ease of use, especially for researchers at any skill level. While "manual" may sound limiting, this approach still significantly enhances the speed and efficiency of qualitative data analysis by allowing ChatGPT's AI capabilities to streamline the identification and organization of codes, patterns, and themes within QualCoder. This method preserves the researcher's full control over their data analysis process, ensuring alignment with research objectives and maintaining the rigor essential to high-quality qualitative research.

1. Manual Import

This approach is ideal for researchers who prefer full control over their data analysis process. After ChatGPT generates codes and suggests themes, researchers manually upload their transcripts into QualCoder. They then apply ChatGPT's coding suggestions within the software, using these recommendations as a guide for their analysis. This method is straightforward and accessible for beginners, allowing the researcher to review, refine, and ensure alignment with their research objectives. The advantage of this approach lies in ChatGPT's ability to speed up the identification of codes, allowing the researcher to focus on applying these codes within QualCoder. After codes are created in QualCoder, the researcher can leverage the software's built-in tools for deeper analysis. ChatGPT can also assist in identifying patterns and themes within the codes, but the researcher manually creates these categories in QualCoder, maintaining full control over the analysis.

2. Semi-Automated Transfer Using CSV or TXT Files

This method is suitable for researchers with basic technical skills who seek to partially automate the transfer process. In this approach, the codes and themes generated by ChatGPT are exported in formats such as TXT or CSV, which can then be imported into QualCoder. While this method speeds up the initial coding setup, the corresponding quotes from the transcripts do not transfer automatically, requiring the researcher to manually associate relevant text with the imported codes. However, the limitation can be mitigated by using QualCoder's memo feature to add notes or contextual information to each code or category. This approach strikes a balance between automation and manual oversight, saving time while ensuring flexibility in the analysis.

3. Fully Automated Import via API Integration:

The most advanced integration method involves using APIs (Application Programming Interfaces) to automate the entire transfer process from ChatGPT to QualCoder. With API connections, codes, themes, and patterns generated by ChatGPT are directly transferred into QualCoder, making it an ideal solution for researchers managing large datasets or seeking maximum efficiency. This method requires more technical expertise, as it involves developing API connections for both platforms. While QualCoder supports the import of REFI-QDA (Rotterdam Exchange Format Initiative) files (22), ChatGPT currently cannot export data in this format. Therefore, a third-party application or another qualitative analysis software may be needed to facilitate the transfer. Though powerful, this fully automated approach falls outside the scope of this book and is recommended for those with advanced technical skills.

Each approach to integrating ChatGPT with QualCoder allows researchers to optimize their workflow while ensuring flexibility and rigor in qualitative data analysis. By selecting the method that best fits their needs and technical abilities, researchers can harness the combined strengths of AI-driven coding and the powerful features of QualCoder.

The following sections elaborate on each of these approaches.

MANUAL IMPORT

The **Manual Import** approach to integrating ChatGPT with QualCoder is well-suited for researchers who want to maintain full control over their data analysis process (Figure 8.1). This method is ideal for those who prefer hands-on interaction with their data, ensuring that every stage of the analysis is carefully curated and that the final output fully aligns with their research objectives. The following sections will break down the key steps involved in using this approach.

Step 1: Generate Codes and Themes with ChatGPT

The process begins by utilizing ChatGPT to help identify initial codes and themes from your qualitative data, such as interview transcripts or focus group discussions. Here's how to get started:

- **Data Input**: Enter your cleaned and formatted transcripts or qualitative text into ChatGPT. You can ask ChatGPT to identify key concepts, phrases, or recurring ideas in the data. Example prompts include:
  "Please identify recurring themes related to workplace stress in this interview transcript."
  "What are the main codes that can be derived from this discussion about remote work productivity?"
- **Code Generation**: Based on your input, ChatGPT will suggest a list of potential codes. These may include specific terms, concepts, or key phrases that repeatedly appear in the data.
- **Theme Identification**: You can take it a step further by asking ChatGPT to help group related codes into broader themes. For example, once ChatGPT has suggested codes like *workload*, *communication barriers*, and *burnout*, you can ask it to help combine these into a larger theme such as *workplace stressors*.

Note: This step accelerates the initial stage of analysis by providing a first draft of codes and themes. However, the researcher should carefully review and refine these suggestions to ensure they fit the context and focus of the research.

Step 2: Upload Transcripts to QualCoder

Once you have generated your codes and themes with ChatGPT, the next step is to manually upload your transcripts into QualCoder for further analysis:

- **Format Transcripts**: Ensure that your transcripts are properly formatted (e.g., cleaned of extraneous information, clearly structured) before importing them into QualCoder. QualCoder accepts plain text, RTF, and other common formats.
- **Import Data**: Open QualCoder and use the "Manage Files" option to upload your transcript files into the project. This makes your raw data available for analysis within the software environment.

Tip: Before proceeding to coding, take some time to explore the built-in tools and options within QualCoder, such as file management, memo creation, and case management, to get familiar with how your data is organized within the software.

Step 3: Create Codes in QualCoder

Now that the transcripts are imported, it is time to create the codes within QualCoder based on ChatGPT's suggestions:

- **Manual Code Creation**: In QualCoder, navigate to the "Code" section and manually create codes that match the ones suggested by ChatGPT. For each code, input a name and a description that clearly defines what it represents. This description helps ensure consistency when applying the codes later in the process.
- **Apply Codes to Text**: After creating your codes, go through the transcript data and manually highlight the relevant text segments that correspond to each code. For example, if ChatGPT suggested the code *workplace stress*, you would identify sections of the transcript that discuss stressful experiences at work and apply the *workplace stress* code to those passages.

This step may take time, but it ensures a hands-on connection with your data, allowing you to review, interpret, and code text based on your deep understanding of the research context.

Step 4: Leverage QualCoder's Tools for Deeper Analysis

Once your data is coded, you can use QualCoder's built-in tools to further analyze your data. This includes:

- **Code Frequency Analysis**: Use the built-in frequency analysis to see how often each code appears within the dataset, helping you quantify the importance or recurrence of specific themes.
- **Co-occurrence Analysis**: Explore how often different codes appear together within the same segments of text. This can reveal relationships between concepts, which is valuable for understanding deeper patterns in the data.
- **Memoing**: In QualCoder, you can add memos or notes to specific codes, sections, or entire files to capture your thoughts, reflections, and emerging insights. This is crucial for keeping track of your interpretive process as the analysis evolves.

Step 5: Refine Themes and Categories Manually

As you progress with coding and begin to see patterns in the data, you can ask ChatGPT to help refine or suggest additional themes. However, the process of grouping codes into broader categories or themes should be manually executed within QualCoder:

- **Theme Development**: Based on your analysis, manually group related codes into categories and themes within QualCoder. For example, if codes like *burnout, heavy workload*, and *long hours* all relate to stress, you might create a theme titled *Workplace Stress* and assign these codes under it.
- **Category Creation**: QualCoder allows you to organize these themes into hierarchies or categories, which helps to structure your final analysis and report. This process is fully manual, giving the researcher full control over the relationships between codes and themes.

Step 6: Finalize the Analysis

At this stage, you have applied ChatGPT's suggestions and manually refined your coding and themes within QualCoder. You can now proceed to the final stages of your analysis:

- **Generate Reports**: Use QualCoder's reporting features to generate summaries of your coded data, including code frequencies, co-occurrence reports, and thematic summaries. These reports can be exported for inclusion in your final analysis or publication.
- **Cross-Analysis**: To ensure rigor, you can conduct cross-analysis by comparing ChatGPT's generated codes with your manually applied codes to validate their relevance and accuracy.
- **Review and Reflect**: As you finalize your analysis, use memos, summaries, and visualizations (like code relationships or word clouds) within QualCoder to reflect on the findings, ensuring that your conclusions are well-supported by the data.

Advantages of the Manual Import Approach

- **Full Control**: This approach gives you complete control over the analysis, ensuring that all codes and themes are appropriately applied to your research context.
- **Hands-on Engagement**: By manually reviewing and coding the data, you remain closely engaged with the qualitative material, allowing for a deeper understanding of the nuances in your transcripts.
- **Flexible and Beginner-Friendly**: This method is accessible to beginners who may not have technical expertise, as it requires no programming skills or complex setup.
- **Customizable**: Even though ChatGPT helps generate initial codes, you can fully customize and refine them within QualCoder to align with your unique research goals.

Figure 8.1. Manual Import approach to integrating ChatGPT with QualCoder

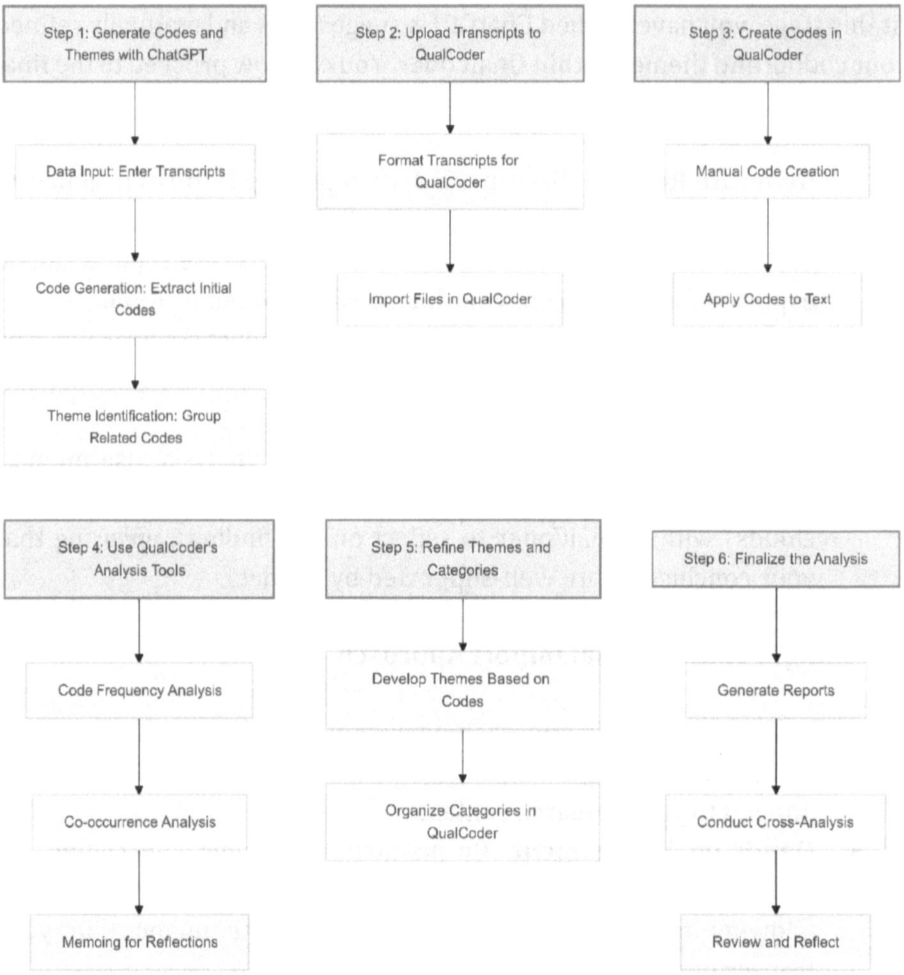

```
Step 1: Generate Codes and          Step 2: Upload Transcripts to        Step 3: Create Codes in
   Themes with ChatGPT                      QualCoder                         QualCoder
           │                                     │                                │
           ▼                                     ▼                                ▼
   Data Input: Enter Transcripts       Format Transcripts for            Manual Code Creation
                                              QualCoder
           │                                     │                                │
           ▼                                     ▼                                ▼
   Code Generation: Extract Initial    Import Files in QualCoder          Apply Codes to Text
              Codes
           │
           ▼
   Theme Identification: Group
         Related Codes


Step 4: Use QualCoder's             Step 5: Refine Themes and          Step 6: Finalize the Analysis
     Analysis Tools                         Categories
           │                                     │                                │
           ▼                                     ▼                                ▼
   Code Frequency Analysis            Develop Themes Based on             Generate Reports
                                              Codes
           │                                     │                                │
           ▼                                     ▼                                ▼
   Co-occurrence Analysis             Organize Categories in             Conduct Cross-Analysis
                                              QualCoder
           │                                                                      │
           ▼                                                                      ▼
   Memoing for Reflections                                                 Review and Reflect
```

SEMI-AUTOMATED TRANSFER USING CSV OR TXT FILES

The **Semi-Automated Transfer** approach provides a balance between automation and manual oversight in the integration of ChatGPT with QualCoder (Figure 8.2). This method is suitable for researchers with basic technical skills who want to automate part of the process while maintaining control over how codes are applied. It involves exporting codes and themes generated by ChatGPT in file formats like CSV or TXT, which are then imported into QualCoder for further analysis. This approach significantly speeds up the initial setup of codes and themes while still requiring manual steps for applying those codes to specific quotes in the data.

Here is a detailed breakdown of each substep involved in this approach.

Step 1: Generate Codes and Themes with ChatGPT

As with the manual method, you begin by using ChatGPT to generate codes and themes from your qualitative data. The steps are as follows:

- **Input Data into ChatGPT**: Provide ChatGPT with your qualitative transcripts or other text-based data and request code suggestions. Example prompts might include: *"Identify key themes in this interview about remote work challenges,"* or
"Generate codes for discussing work-life balance in the following text."
- **Code and Theme Generation**: ChatGPT will provide you with a list of codes and potential themes based on the content. For instance, it might generate codes such as *remote collaboration*, *productivity issues*, and *employee well-being*.
- **Review and Refine**: Before moving forward, review ChatGPT's suggestions and make any necessary adjustments. This ensures that the codes are relevant and aligned with your research questions.

Step 2: Export Codes and Themes in CSV or TXT Format

Once ChatGPT has generated your codes and themes, the next step is to export these into a format that can be imported into QualCoder.

- **Copy Codes and Themes**: Using ChatGPT's output, copy the list of codes and themes into a structured format. You can do this manually by copying and pasting the codes and categories from the chat window, or by generating a more organized list in ChatGPT using prompts like *"Generate a CSV-friendly list of codes and themes."*
- **Create a CSV or TXT File**: Save the codes and themes in a CSV or TXT file format. A CSV file is especially useful for structuring data in rows and columns, which QualCoder can read more easily. The basic structure of the CSV file should look something like this:

Table 8.1. Organizing Codes for Import into QualCoder

Code	Theme	Description
Remote Collaboration	Team Dynamics	Issues with remote communication
Productivity Issues	Work Performance	Difficulties maintaining productivity
Employee Well-being	Emotional Health	Concerns about work-life balance

This format ensures that your codes are organized, and it allows for easy import into QualCoder (Table 8.1).

Step 3: Import Codes and Themes into QualCoder

Now that your codes are saved in CSV or TXT format, you can import them into QualCoder, saving time in manually setting up the code framework.

- **Open QualCoder**: Launch the QualCoder software and open your project.
- **Import Codes**: Navigate to the "Code" section within QualCoder and use the "Import" function to upload the CSV or TXT file containing your codes. QualCoder will automatically create the codes listed in the file and associate them with the project.
- **Review the Imported Codes**: Once imported, review the codes within QualCoder to ensure that they have been correctly mapped. Verify that each code has its associated description and theme as intended. This review is crucial to avoid any formatting issues that may have occurred during the import.

Step 4: Manually Associate Codes with Text Segments

One limitation of this approach is that while codes and themes are imported automatically, the corresponding quotes or text segments from the transcripts are not. Therefore, the researcher must manually associate the imported codes with relevant portions of the text.

- **Identify Relevant Text Segments**: Open your transcript in QualCoder and manually highlight sections of text that correspond to each code. For example, for the code *Employee Well-being*, identify sections where participants discuss their emotional health and apply the appropriate code to those sections.
- **Apply Codes**: As you highlight relevant text segments, apply the corresponding codes that were imported into QualCoder. This step remains manual but ensures that you carefully review and interpret the data, maintaining a hands-on approach to coding.

Step 5: Use Memos for Contextual Notes

To mitigate the limitation of not having the quotes automatically linked to the imported codes, you can use QualCoder's **Memo** feature to add contextual notes for each code or theme.

- **Create Memos**: For each code, add memos that explain its context or any insights you have gathered from the data. For instance, if *Remote Collaboration* is a code, you might create a memo explaining that this code covers discussions about challenges in remote teamwork and communication.
- **Link Memos to Codes**: Attach memos to specific codes or categories, providing additional context for each one. This feature helps keep track of insights and reflections that are not automatically generated during the import process.

Tip: You can use memos not only to document your thinking but also to store any additional information from ChatGPT that might provide further context for specific codes or themes.

Step 6: Refine Codes and Themes Within QualCoder

After associating codes with the appropriate text segments, you can refine and organize them into broader categories or themes.

- **Develop Categories**: Manually group related codes into categories or themes within QualCoder. For instance, if you have multiple codes related to team communication and collaboration, you might group them under a theme like *Remote Work Challenges.*
- **Hierarchical Structuring**: QualCoder allows you to create hierarchies of codes, which can help in structuring your final analysis. For example, under the theme *Employee Well-being*, you might have sub-codes such as *Stress*, *Work-life Balance*, and *Mental Health.*

Step 7: Finalize and Analyze Data

Once your codes are applied and refined, you can utilize QualCoder's built-in tools for further analysis.

- **Generate Reports**: QualCoder allows you to generate code frequency reports, which summarize how often each code appears in the data. You can export these reports for use in presentations or publications.
- **Visualize Relationships**: Use QualCoder's visual tools, such as code relationship diagrams and word clouds, to identify patterns and connections between codes.
- **Cross-Check Codes**: To ensure thorough analysis, you can compare the codes generated by ChatGPT with the ones you have manually refined. This helps validate the accuracy and relevance of ChatGPT's output.

Advantages of the Semi-Automated Transfer Approach

- **Saves Time**: This method speeds up the process of creating codes and themes by automating the import process through CSV or TXT files. You do not need to manually input each code or theme into QualCoder, saving considerable time in the initial setup phase.

- **Flexibility**: While some automation is involved, this approach still allows for manual refinement and review of the data, ensuring that the researcher maintains control over the analysis.
- **Memo Feature**: Using memos to supplement the codes adds flexibility, allowing researchers to document their thoughts and reflections even though quotes are not automatically transferred.
- **Beginner-Friendly with Basic Technical Skills**: The process only requires basic technical skills like handling CSV files, making it accessible to researchers who are comfortable using spreadsheets but may not have programming knowledge.

Figure 8.2. Semi-Automated Transfer approach from ChatGPT to QualCoder

FULLY AUTOMATED IMPORT VIA API INTEGRATION

The **Fully Automated Import via API Integration** is the most advanced approach for integrating ChatGPT with QualCoder (Figure 8.3). It automates the entire process of transferring codes, themes, and patterns generated by ChatGPT directly into QualCoder through the use of APIs. This method is ideal for researchers dealing with large datasets who require maximum efficiency. However, it requires a solid understanding of programming and API development to set up the necessary connections between ChatGPT and QualCoder.

In this section, we will outline the key steps involved in setting up a fully automated system using APIs and discuss the advantages and challenges associated with this method.

Step 1: Understanding APIs and Setting Up ChatGPT and QualCoder for Integration

Before diving into the technical setup, it is important to understand the role of APIs in facilitating the automated transfer of data between ChatGPT and QualCoder.

- **What is an API?**: APIs allow two software platforms to communicate and exchange data automatically. In this context, the API will enable ChatGPT to send the generated codes, themes, and patterns directly to QualCoder without manual intervention.
- **Setting Up OpenAI's API (ChatGPT)**: You will need to create an account with OpenAI and obtain an API key for accessing ChatGPT. The API key allows you to send and retrieve data from ChatGPT programmatically.
    - **Obtain an API Key**: Sign up on OpenAI's platform and navigate to the API section. From there, you can generate an API key, which will be used in your code to access ChatGPT's services.
    - **API Documentation**: Review the OpenAI API documentation to understand the available methods for sending data to ChatGPT and receiving its output. This includes methods for generating codes and themes from textual data.

- **Setting Up API Access for QualCoder**: QualCoder does not currently offer a native API, but it supports the import of REFI-QDA files (a standardized format for qualitative data). To achieve full automation, you will need to either develop a custom API for QualCoder or use a third-party tool that supports API-based data transfer.
    - **Develop Custom API Solutions**: If you have advanced programming skills, you could create a custom script or middleware that acts as an API for QualCoder, converting ChatGPT's output into a format that QualCoder can interpret and import.
    - **Third-Party Applications**: If direct API integration with QualCoder is challenging, consider using third-party software that supports APIs and REFI-QDA file formats to act as a bridge between ChatGPT and QualCoder.

Step 2: Automating Data Transfer from ChatGPT to QualCoder

Once API access is set up, the next step is to automate the transfer of codes and themes from ChatGPT into QualCoder.

- **Sending Data to ChatGPT via API**: First, programmatically send your qualitative data (e.g., interview transcripts) to ChatGPT via its API. This can be done by writing a script that feeds text data into ChatGPT and requests code and theme generation. Example API request:

```python
Copy code
import openai

openai.api_key = 'your-api-key-here'

response = openai.Completion.create(
    engine="gpt-4",
    prompt="Identify key codes and themes from this interview transcript: [your transcript text here]",
    max_tokens=1000
```

)

codes_and_themes = response['choices'][0]['text']

The above script will send your transcript to ChatGPT and receive a response containing the suggested codes and themes.

- **Converting Output into a Structured Format**: Once ChatGPT returns the codes and themes, format the output into a structure that can be automatically processed by QualCoder. This typically involves saving the output as a CSV or JSON file. For example, a simple conversion could output the codes into a CSV file like this:

```
python
Copy code
import csv

with open('codes.csv', mode='w', newline='') as file:
    writer = csv.writer(file)
    writer.writerow(["Code", "Theme", "Description"])
    for code, theme, description in codes_and_themes:
        writer.writerow([code, theme, description])
```

This script creates a CSV file that can later be imported into QualCoder.

- **Automated Import into QualCoder**: If you are developing an API for QualCoder or using a third-party tool, the next step is to set up a system where the CSV or JSON file generated by ChatGPT is automatically imported into QualCoder. This can be done using custom middleware that reads the file and inserts the codes and themes into QualCoder's database or project file structure.

Step 3: Handling the REFI-QDA Format for QualCoder Integration

Since QualCoder supports the REFI-QDA standard, you may need to convert the output from ChatGPT into this format to fully automate the import process.

- **What is REFI-QDA?**: REFI-QDA is a standard format used by qualitative analysis software to exchange data (e.g., codes, themes, and transcript information) between platforms. While ChatGPT does not natively support REFI-QDA, you can create a script that formats ChatGPT's output into this format.
  - **Conversion Script**: Write a script that takes the output from ChatGPT and converts it into the REFI-QDA format. This may involve mapping the codes and themes from ChatGPT into the XML-based structure of REFI-QDA.
  - **Third-Party Tools**: Alternatively, use third-party applications like NVivo or ATLAS.ti, which support both API integrations and REFI-QDA imports, to act as intermediaries in the process. These tools can help bridge the gap between ChatGPT's output and QualCoder's requirements.

Step 4: Running the Automated Workflow

With the APIs and data conversion processes in place, you can now automate the entire workflow. Here's a simplified outline of the process:

1. **Data Submission**: The researcher uploads the qualitative data (transcripts, notes, etc.) into the automation script, which sends it to ChatGPT via the API.
2. **Code and Theme Generation**: ChatGPT processes the data and returns a structured list of codes and themes.
3. **File Conversion**: The output is automatically converted into a CSV, JSON, or REFI-QDA format.
4. **Automatic Import into QualCoder**: The codes and themes are transferred directly into QualCoder's project environment through either a custom API connection or a third-party software intermediary.
5. **Code Application**: In the most advanced setups, the system may also attempt to apply the codes directly to relevant text segments in QualCoder, although this feature may still require some manual intervention depending on the complexity of the data.

Step 5: Reviewing and Finalizing the Analysis

After the fully automated import process, it's essential to review the results to ensure accuracy and relevance.

- **Review Codes and Themes**: Although the transfer is automated, it's important to manually review the codes and themes within QualCoder to verify that they align with your research objectives.
- **Refine as Needed**: Make any necessary adjustments to the codes, group related codes into categories, and create memos to document your analytical process.
- **Apply QualCoder Tools**: Once the data is imported, you can utilize all of QualCoder's built-in analysis tools, such as code frequency analysis, co-occurrence tables, and visualization features, to conduct deeper analysis.

Challenges and Considerations for Fully Automated Integration

While the fully automated approach offers efficiency and scalability, it comes with certain challenges:

- **Technical Expertise**: This approach requires significant programming knowledge to set up the API connections and develop scripts for automating the data transfer process. Researchers without coding experience may find this approach difficult to implement.
- **Limited Format Compatibility**: ChatGPT does not currently support the REFI-QDA format, which is essential for QualCoder. Therefore, additional tools or custom scripts are needed to convert ChatGPT's output into a compatible format.
- **Debugging and Maintenance**: API connections and scripts may require ongoing debugging and maintenance, especially if there are updates to either ChatGPT or QualCoder. Researchers need to ensure their automated workflows remain functional as software platforms evolve.

Advantages of the Fully Automated Approach

- **Efficiency with Large Datasets**: This method is particularly useful for researchers handling large volumes of qualitative data, as it eliminates the need for manual coding and data entry.
- **Seamless Integration**: When set up correctly, this approach offers a seamless, end-to-end integration between ChatGPT and QualCoder, allowing researchers to focus more on analysis rather than on administrative tasks.
- **Scalability**: This approach can be scaled to handle multiple projects or large datasets, making it an ideal solution for research teams or organizations with extensive data analysis needs.

Figure 8.3. Fully Automated Import from ChatGPT to QualCoder via API Integration

Chapter Key Points:

- *Integrating ChatGPT with QualCoder enhances qualitative data analysis by combining AI capabilities with a powerful open-source tool for efficient research.*
- *There are three integration approaches—Manual Import, Semi-Automated Transfer, and Fully Automated Import via API Integration—each varying in automation level and technical expertise required.*
- *The Manual Import method involves manually applying ChatGPT-generated codes in QualCoder, offering full control and alignment with research objectives.*
- *The Semi-Automated Transfer method uses CSV or TXT files to import ChatGPT codes into QualCoder, saving setup time but requiring manual association of codes with text segments.*
- *The Fully Automated Import via API Integration method automates code transfer from ChatGPT to QualCoder using APIs, ideal for large datasets but requiring advanced technical skills and possibly additional tools.*
- *Each approach allows researchers to optimize their workflow based on their needs and skills while ensuring rigorous qualitative analysis.*
- *Challenges with the Fully Automated approach include technical complexity and compatibility issues, necessitating programming expertise and potential use of third-party tools.*
- *Selecting the appropriate integration method enables leveraging AI-driven coding with QualCoder's features, enhancing efficiency without compromising research quality.*

Chapter 9.

PREPARING YOUR RESEARCH ENVIRONMENT FOR AI INTEGRATION

ACCESSING AND CONFIGURING CHATGPT

To effectively use ChatGPT for qualitative data analysis, it is crucial to understand the different approaches to accessing and configuring the tool. Broadly, there are two main ways to utilize ChatGPT for qualitative analysis:

1. **Using the Default GPT Models** (via OpenAI's web interface or API)
2. **Building Custom GPTs** tailored specifically for qualitative analysis.

Each approach offers unique advantages and limitations, and your choice will depend on the complexity of your project, your technical skills, and the specific needs of your research (Table 9.1).

1. Using Default GPT Models for Qualitative Analysis

The most common and straightforward approach is to use the **default GPT models** (such as GPT-4) directly through OpenAI's web interface or via API integration (23). This approach offers ease of access and requires little to no technical expertise.

- **Accessing ChatGPT**: You can easily access the GPT models through the OpenAI web interface (chat.openai.com) or integrate the tool into your workflow using the OpenAI API. This allows you to input qualitative data, such as interview transcripts, and receive generated themes, codes, or even summaries directly from ChatGPT.
- **Subscription Plans**: If you are conducting complex analyses or need to process large amounts of qualitative data, consider using **ChatGPT Plus**, which provides access to the more advanced **GPT-4**

model. GPT-4 offers better language comprehension, increased accuracy, and can handle more complex and nuanced inputs compared to previous versions.
- **Configuring for Research Needs**:
  - **Prompt Crafting**: When using the default models, your success will largely depend on how well you craft your prompts. For example, you can instruct ChatGPT to "identify key themes" in a set of responses, or you can ask it to "generate initial codes based on this transcript." You will need to ensure that the prompts are clear and specific to yield useful results.

Pros of Using Default GPT Models:

- **Ease of Access**: The default models are readily available through a simple user interface, requiring minimal setup or technical knowledge.
- **Quick and Efficient**: You can instantly start analyzing your qualitative data without needing to build or configure specialized models.
- **Widely Applicable**: Default GPT models are versatile and can handle a wide range of qualitative tasks, from summarizing to thematic analysis.

Cons of Using Default GPT Models:

- **Generalized Output**: Since the model is not customized for your specific data or analysis needs, the results might be broad or lack the depth required for complex qualitative studies.
- **Limited Customization**: You cannot fine-tune the default models to specialize in qualitative research, which could limit their effectiveness in identifying very niche or specific themes.

2. Building Custom GPTs for Qualitative Analysis

For researchers seeking tailored support in their qualitative analysis, OpenAI offers the ability to create **Custom GPTs** using its inbuilt tools. This feature allows you to build and customize GPTs specifically for your project, without needing to use APIs or complex programming. You can create a

Custom GPT by providing specific instructions and examples, ensuring that the model is more focused on the unique themes and requirements of your qualitative research.

What Are Custom GPTs?

Custom GPTs are personalized versions of the GPT models that can be adapted to meet your specific research needs. Through OpenAI's user-friendly interface, you can customize how the GPT behaves by inputting your data and crafting detailed instructions for your qualitative study. For example, if you are analyzing interview data related to mental health, you can create a Custom GPT that focuses on identifying concepts such as stress, coping mechanisms, or emotional well-being based on your instructions and examples.

Building a Custom GPT

- **Defining Your Instructions and Examples**: Begin by outlining the specific tasks you want the Custom GPT to perform. You can guide the model by providing detailed instructions and sample questions that help it focus on the most relevant aspects of your research.
- **Customizing the GPT**: OpenAI's inbuilt tools allow you to customize GPT behavior without the need for programming or APIs. You can define the model's tone, clarify its role in your research, and upload sample data to give it context for analysis. For instance, you can instruct it to focus on recognizing themes related to work-life balance, emotional stress, or social interaction.
- **Testing and Refining**: Once you have created a Custom GPT, you can test it by asking specific research-related queries. Based on its responses, you can further refine the instructions and behavior to ensure it produces accurate and insightful outputs that align with your research objectives.

Pros of Building Custom GPTs

- **Tailored to Your Research**: Custom GPTs can be designed to focus on your specific qualitative data, producing more relevant codes and themes that are aligned with your research objectives.

- **Greater Ease of Use**: Unlike the technical complexity of building a model from scratch with APIs, OpenAI's inbuilt Custom GPT creation tool is user-friendly and does not require any programming knowledge. This makes it accessible to researchers with varying levels of technical expertise.
- **Quick Customization**: Custom GPTs can be set up relatively quickly, enabling you to adjust their behavior and focus without extensive setup, making them ideal for time-sensitive projects.

Cons of Building Custom GPTs

- **Limited to Pre-Defined Capabilities**: While the customization options are flexible, you are still working within the existing framework of GPT, which may limit more advanced or highly specific requirements compared to custom models trained with APIs.
- **Requires Ongoing Refinement**: Although easy to set up, Custom GPTs may still require iterative refinement based on testing and feedback to ensure they are aligned with your specific research needs.
- **Potential Costs**: While using Custom GPTs does not involve the higher costs of API-driven fine-tuning, there may still be usage fees, depending on your data volume and customization needs.

By leveraging OpenAI's inbuilt Custom GPT tools, researchers can easily create specialized models that suit their qualitative research needs without needing extensive technical expertise or resources. This makes Custom GPTs an ideal solution for tailoring AI-powered analysis to your project's unique themes and questions.

CHOOSING THE RIGHT APPROACH

Deciding between using the **default GPT models** or creating **custom GPTs** depends on your specific research needs, timeline, and resources. If you are looking for quick, general insights and do not require deep customization, the default models will likely suffice. However, if your project involves complex qualitative analysis with specific themes or language nuances, building a custom GPT may offer greater precision and relevance.

For most general qualitative research projects, starting with the default models and crafting specific prompts will be the most practical and efficient option. As your research evolves, you may consider transitioning to custom GPTs if you need more specialized analysis.

Note: GPT models continue to be updated with new functionalities, and improvements are made over time. For the most recent features and versions, it is advisable to check the OpenAI website regularly to stay informed on the latest advancements and capabilities of GPT models.

Table 9.1. Approaches to Accessing and Configuring ChatGPT

Aspect	Using default gpt models	Building custom gpts (using openai's inbuilt tool)
Definition	Using standard GPT models like GPT-4 via OpenAI's web interface for qualitative tasks.	Creating specialized GPTs customized to your research needs through OpenAI's built-in GPT customization feature, without requiring APIs.
Access method	- Access via OpenAI's web interface. - Requires minimal setup or technical knowledge.	- Access via OpenAI's web interface (Custom GPT feature). - No API or programming needed; built directly in OpenAI's user interface.
Configuration	- Craft clear and specific prompts for desired results. - Adjust temperature and tone settings via the interface for controlled outputs.	- Provide detailed instructions and examples to guide the model. - Customize the behavior of the GPT without the need for technical coding skills.
Best use case	- Quick, efficient access for general qualitative tasks like summarizing, identifying themes, and generating initial codes. - Suitable for straightforward research projects.	- Ideal for research requiring more specific focus or themes relevant to your study (e.g., exploring stress management, specific cultural patterns). - Great for researchers who want more tailored insights without coding or API setup.
Pros	- **Ease of Access**: Simple interface, ready-to-use. - **Quick Results**: Immediate access to qualitative analysis.	- **Tailored to Research**: Easily customizable to produce more accurate, project-specific outputs. - **No Technical Expertise Required**: No programming or APIs needed.

Cons	- **Versatile**: Can handle a wide range of qualitative tasks. - **Generalized Output**: May lack depth for highly specialized qualitative analysis. - **Limited Customization**: Only basic adjustments to outputs and behavior are available.	- **Deeper Insights**: Provides nuanced data based on specific instructions. - **Limited to Pre-Defined Frameworks**: While highly customizable, the features are still confined to the broader GPT model's capabilities. - **Ongoing Refinement**: May require iterative adjustments to perfect the outputs.
Subscription/ cost	- ChatGPT Plus subscription gives access to GPT-4 with better performance for complex tasks.	- Usage fees depend on data volume and customization, but no API costs are involved.
Example use case	- "Identify key themes related to work-life balance in this interview transcript."	- "Analyze the specific emotional responses of participants in this focus group about stress, based on tailored instructions provided."

This table provides a clear comparison between using **default GPT models** and **building custom GPTs** for qualitative research, helping researchers determine the best approach based on their project needs, technical skills, and resources.

DATA PREPARATION: CLEANING AND FORMATTING QUALITATIVE DATA

Once you have accessed ChatGPT and decided on either using the **default models** or **custom GPTs**, the next step is preparing your qualitative data for analysis (Figure 9.1). Raw qualitative data—such as interview transcripts, focus group discussions, or observational notes—often needs cleaning and formatting to make it compatible with AI-assisted analysis. Proper data preparation ensures that ChatGPT delivers accurate, meaningful, and relevant results during your thematic analysis.

Steps for Data Cleaning and Formatting:

TRANSCRIPTION ACCURACY

If your data includes interviews, focus groups, or other audio/video recordings, it is essential to have accurate and clear transcripts. While there are several transcription tools available, like Otter.ai, Rev, or Descript, always manually review the transcripts to correct any errors. ChatGPT's performance in qualitative analysis heavily depends on the quality of the input. Inaccuracies in the transcript—such as misheard words or missed sentences—can lead to incorrect coding or misidentified themes.

- **Using Automated Transcription Tools**: Automated transcription tools are an efficient way to quickly convert audio data into text. However, no matter how advanced, AI-driven transcription tools are not perfect. Misinterpretation of jargon, speaker changes, and background noise can cause inaccuracies, which is why manual review is critical.
- **Manual Review**: After generating the transcript, review it thoroughly to ensure that important nuances are not lost or misrepresented. This step is vital when analyzing complex interviews with specific terminology or context.

FORMATTING FOR AI READABILITY

To ensure that ChatGPT can effectively analyze your data, it is important to format the transcripts in a way that makes them easy for AI to process. Formatting plays a key role in helping ChatGPT recognize the structure of your data and respond accurately during analysis.

- **Remove Extraneous Information**: Unnecessary information, such as time stamps, filler words ("um," "you know," "like"), or repetitive interviewer prompts, should be removed or cleaned up. These elements can confuse ChatGPT and lead to irrelevant codes or themes.
    - Example:
        - Raw: "Um, yeah, so I guess… like, when we use VR, it's, um, interesting but also a bit, you know, draining."
        - Cleaned: "When we use VR, it's interesting but also a bit draining."
- **Standardize Interview Structures**: If you are analyzing multiple interviews or discussions, ensure that all transcripts follow a consistent structure. Clearly distinguish between the interviewer's questions and the participant's responses using headings such as **"Interviewer"** and **"Participant"**. This allows ChatGPT to better understand the flow of the conversation and to focus on the relevant content in its analysis.
    - Example Format:
        - **Interviewer**: "How has using VR affected your productivity?"
        - **Participant**: "It's helped for creative tasks but has been a challenge for regular meetings."
- **Tag Key Segments**: Use labels or tags to mark key areas in your transcripts. This makes it easier for ChatGPT to focus on specific sections during analysis, such as responses directly related to productivity or well-being. You can introduce sections like "Team Collaboration" or "Emotional Well-being" for clarity and focus.

BREAKING DOWN LARGE DATA SETS

ChatGPT has token limits, meaning it can only process a certain amount of text in a single query. If your interview transcripts or datasets are lengthy, you will need to break them into smaller, manageable chunks. This step is essential to avoid overloading the model and losing context during analysis.

- **Chunking Large Transcripts**: If an interview is long or has multiple sections, break it down by responses to each specific question or by theme (e.g., **"Productivity"**, **"Team Interaction"**, **"Well-being"**). By

keeping the sections logically structured, you allow ChatGPT to maintain the necessary context across smaller batches of data.
- o Example:
    - Instead of submitting an entire 3,000-word interview, break it into sections like:
        - Part 1: Responses about productivity (1,000 words)
        - Part 2: Responses about team interaction (1,000 words)
        - Part 3: Responses about well-being (1,000 words)

This approach ensures that the AI focuses on specific aspects of the data, generating more accurate and theme-specific insights.

DATA CLEANING FOR THEMES

Once your transcripts are formatted and divided into manageable chunks, it is important to clean your data with your predefined themes in mind. For deductive thematic analysis, you will have already identified key themes or categories from your theoretical framework. Preparing your data in line with these themes helps ChatGPT focus on the right content and produce relevant codes or insights.

- **Pre-Identifying Themes**: Highlight sections of the transcript that correspond to your key areas of interest. For example, if your predefined themes include "communication," "productivity," and "emotional well-being," make sure that relevant segments of the transcript are organized accordingly. This can be done manually or through text markers, like:
    - o **Example**:
        - **[Theme: Communication]**
        - **Participant**: "VR has definitely made communication with the team more dynamic, but there's also some awkwardness due to the virtual setting."
        - **[Theme: Productivity]**

- **Participant**: "I feel more productive when I use VR for creative tasks, but routine meetings feel a bit cumbersome."
- **Using Tags and Comments**: If your transcripts are stored in a text editor or qualitative analysis software, use comments or tagging features to identify important sections for analysis. When feeding these into ChatGPT, you can use these markers to ask the AI to focus on specific aspects, ensuring that it aligns with your research themes.

Example Workflow for Data Cleaning:

1. **Transcribe**: Use automated tools for initial transcription, followed by manual review for accuracy.
2. **Clean**: Remove filler words, time stamps, and extraneous information.
3. **Format**: Ensure consistency by clearly labeling the interviewer's questions and the participant's responses, and introduce sections with relevant tags for specific themes.
4. **Chunk**: Divide long interviews into smaller, logically coherent sections to accommodate ChatGPT's token limits.
5. **Pre-Tag Themes**: Identify and label important segments of your transcripts based on the predefined themes of your study, making it easier for ChatGPT to generate relevant codes.

By properly cleaning and formatting your data, you ensure that ChatGPT can process your qualitative transcripts accurately and generate meaningful codes and themes. The next step involves guiding ChatGPT through the analysis process using these well-prepared transcripts and effective prompts.

Figure 9.1. Steps for Data Cleaning and Formatting for Qualitative Analysis with ChatGPT

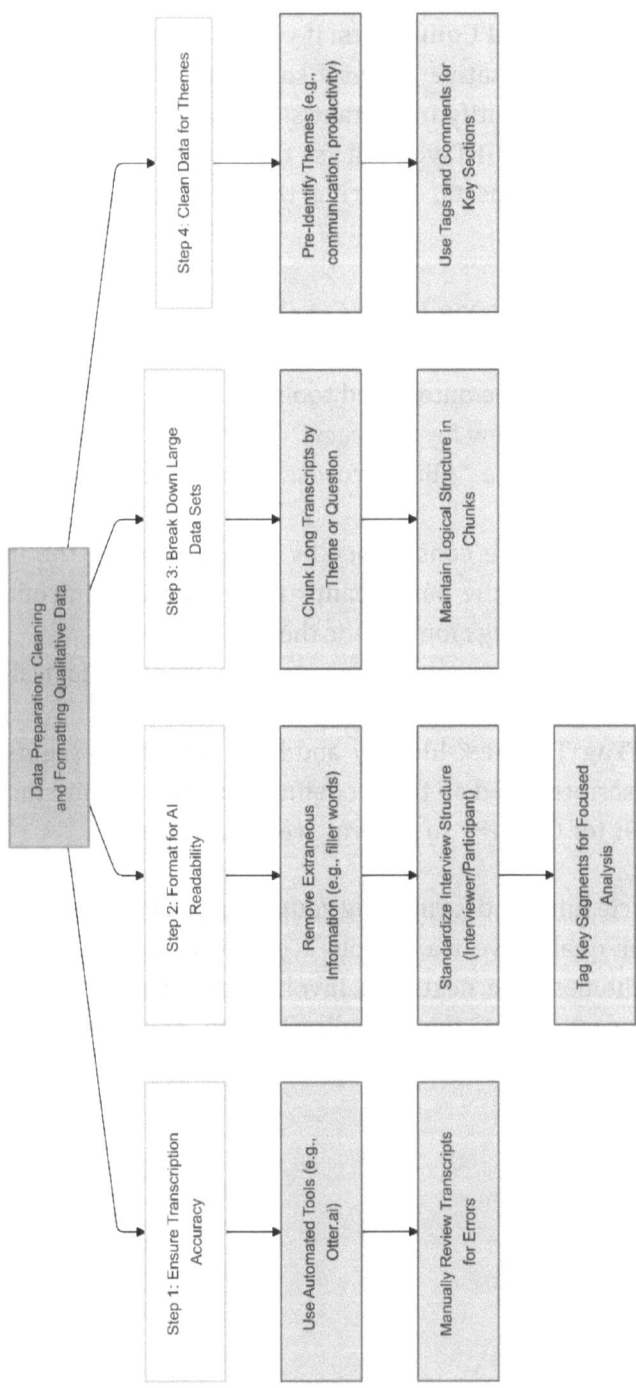

Ensuring Data Privacy and Compliance

Qualitative research often deals with sensitive and personal data, particularly when it involves interviews, focus groups, or any form of data collection that captures individual experiences and insights. When using ChatGPT for qualitative research, ensuring data privacy and compliance with ethical standards becomes even more critical. AI tools like ChatGPT process large amounts of text data, and it is essential to safeguard the privacy of research participants while adhering to legal and ethical requirements.

Here are the key considerations for ensuring data privacy and compliance in AI-assisted qualitative research:

ANONYMIZING DATA

Anonymization is one of the most important steps when preparing qualitative data for analysis with ChatGPT. This involves removing any personal identifiers from the transcripts to protect participants' privacy and ensure confidentiality.

Steps for Anonymizing Data:

- **Remove Identifying Information**: Before feeding your interview transcripts into ChatGPT, make sure all personal identifiers, such as names, job titles, company details, and geographical locations, are removed or replaced with generic labels. This minimizes the risk of identifying individuals from the data.
    - Example:
        - Instead of "John Smith from ABC Corporation," use "Participant 1 from Company A."
        - For geographical details, replace specific cities or locations with broader terms, such as "Region X" instead of "New York City."
- **Pseudonymization**: Assign participants unique pseudonyms or codes (e.g., **"Participant 1"**, **"Team Member A"**) to distinguish them throughout the analysis without revealing their identities. This allows you to maintain clarity in the data while protecting anonymity.

Anonymizing your data before using ChatGPT is crucial for ensuring that sensitive information is not exposed or mishandled during the analysis process.

INFORMED CONSENT

Informed consent is a fundamental ethical principle in research, especially when using AI tools like ChatGPT. Participants must be fully aware of how their data will be used, including its processing by AI systems, and must give their explicit consent before data collection begins.

Steps for Informed Consent:

- **Transparency about AI Use**: Ensure that your consent forms clearly outline how AI, particularly ChatGPT, will be used in the analysis process. Explain that their qualitative data (e.g., interview transcripts) will be processed by an AI system to generate insights or themes, and clarify what this means in layman's terms. Participants should understand how their data will be used, stored, and protected.
- **Consent for External Systems**: Since ChatGPT is an external AI tool, it is essential to inform participants that their data may be processed by a third-party platform (OpenAI in this case). Participants should be made aware of any potential data transfer outside of the primary research institution.
- **Voluntary Participation**: Reaffirm that participation is voluntary and that participants can withdraw at any time without penalty. Provide clear instructions on how they can revoke their consent if they decide not to participate further.

By obtaining informed consent that explicitly covers AI usage, you ensure that participants are fully aware of their rights and the scope of the study.

COMPLIANCE WITH DATA PROTECTION LAWS

When conducting research with sensitive qualitative data, you must comply with relevant data protection laws that govern the storage, processing, and use of personal data. These laws vary by region, and it is important to be familiar with the ones that apply to your research.

Key Data Protection Laws:

- **GDPR (General Data Protection Regulation)**: If your research involves participants in Europe, you need to comply with GDPR, which governs how personal data must be handled and protected. Under GDPR, personal data must be anonymized, securely stored, and processed with explicit consent.
- **CCPA (California Consumer Privacy Act)**: For research involving participants in California, CCPA outlines data privacy rights, including the right to know how personal data is being used and the right to have data deleted upon request.
- **Other Regional Laws**: Depending on the location of your participants, familiarize yourself with local data protection regulations, such as **PIPEDA** in Canada, **HIPAA** (for health-related research) in the U.S., or **Australian Privacy Principles (APPs)**.

Ensure that your use of ChatGPT complies with these legal frameworks. This includes informing participants about their rights under these laws, such as the right to access their data or request its deletion.

STORAGE AND DATA SECURITY

When working with sensitive qualitative data, it is vital to ensure that the data is securely stored and protected throughout the research process, from initial data collection to the analysis phase. Proper storage and security measures protect against unauthorized access or breaches, especially when using cloud-based services or external tools like ChatGPT.

Best Practices for Data Storage and Security:

- **Encryption**: Store all data, including transcripts and analysis outputs, in secure, encrypted databases. This ensures that even if the data is intercepted, it cannot be read or accessed without the decryption key.
- **Cloud Services Compliance**: If you are using cloud-based services to store your data, ensure that these services are compliant with data protection regulations like GDPR or CCPA. Cloud providers should offer robust encryption and access controls.

- **Access Controls**: Limit access to sensitive data to only those who are directly involved in the research. Ensure that all users who can access the data use strong passwords and two-factor authentication (2FA) to prevent unauthorized entry.
- **Backup Procedures**: Regularly back up your data to prevent data loss in case of system failures. However, make sure that backup copies are also encrypted and stored securely.

OpenAI's Data Usage Policies

When using ChatGPT for qualitative analysis, it is important to understand OpenAI's data usage policies. As of the time of writing, OpenAI does not use data sent to its API for training purposes unless the user consents to it. However, this policy may change, so it is essential to stay up to date with OpenAI's terms of service.

What to Review:

- **Data Retention Policies**: Check how long OpenAI retains data submitted via the API and what steps are taken to ensure data privacy. Ideally, you want to ensure that your data is not stored indefinitely and is deleted after the analysis is complete.
- **Data Processing Practices**: Review OpenAI's terms to understand how your data is processed, whether for analysis, storage, or usage in model training. You can choose to opt out of data usage for future model training if this option is available.

By staying informed about OpenAI's data policies, you ensure that your research aligns with your ethical standards and complies with participant consent regarding data usage.

Transparency and Ethical Reporting

When publishing or presenting your research, transparency about how ChatGPT was used is essential. Ethical reporting includes clearly documenting the role that AI played in the data analysis process and addressing any limitations or potential biases that AI may have introduced into the analysis.

Steps for Transparent Reporting:

- **Document ChatGPT's Role**: Clearly explain how ChatGPT was used in the research process—whether for generating codes, identifying themes, or analyzing patterns. This should be part of your methodology section, where you describe the tools and processes used for analysis.

- **Acknowledge Limitations**: AI models like ChatGPT may introduce biases or oversights. Be upfront about these limitations and how you mitigated them, such as reviewing and refining AI-generated codes with human oversight. It is important to acknowledge that while AI can aid in analysis, the final interpretation should still rest with the researcher.
- **Maintain Ethical Boundaries**: Ensure that your analysis adheres to ethical standards, such as fairness, accuracy, and respect for participant data. Do not overstate AI's capabilities, and make sure that human judgment and expertise remain central to the research findings.

INTERVIEW TRANSCRIPT EXAMPLE

The following interview example focuses on how employees and teams have been affected by the introduction of VR in their remote work environments. This could involve using VR for virtual meetings, collaborative projects, and even social interaction in a virtual workspace.

Given the novelty of VR in professional settings, participants could be remote employees or managers who have been experimenting with VR tools to enhance collaboration, communication, and team building. The goal of the interview is to capture their unique experiences and the challenges and opportunities they have encountered.

"The Impact of Virtual Reality (VR) on Team Collaboration in Remote Work Environments"

Rationale:

The rise of remote work has accelerated the use of new technologies, including Virtual Reality (VR), in the workplace. However, the effects of using VR for remote team collaboration have not been extensively studied. This topic is relatively new and allows us to explore how immersive technologies like VR are shaping the way teams interact, collaborate, and communicate in virtual spaces. With VR gaining traction for remote meetings, virtual brainstorming sessions, and training programs, this subject offers a fresh perspective on how cutting-edge technology is influencing workplace dynamics, making it an ideal subject for qualitative analysis.

DEDUCTIVE THEMATIC ANALYSIS

Deductive thematic analysis is well-suited for exploring a novel and evolving topic like VR in remote work environments. In this case, the analysis will be guided by predefined themes based on existing theories of team collaboration and technology adoption in the workplace. By applying these pre-defined themes to interview data, we can explore how the integration of VR is impacting areas such as communication, productivity, interpersonal relationships, and overall team cohesion in a virtual work environment.

Our focus will be on how ChatGPT can assist in identifying key themes and recurring patterns from interview transcripts, aligning these with the pre-established categories, and helping researchers explore the broader implications of VR for team dynamics.

This example allows us to investigate underexplored areas of research, such as whether VR helps bridge the gap between geographically dispersed teams, fosters better communication, or even creates new types of workplace stress or digital fatigue.

STRUCTURE OF THE INTERVIEW

The semi-structured interview design will focus on core areas related to VR and its impact on remote team collaboration. This structure allows for detailed, qualitative responses while ensuring the interview stays aligned with predefined themes.

Interview Structure:

1. **Introduction and General Questions**:
    - "How long have you been using Virtual Reality (VR) tools in your remote work environment?"
    - "What are the main tasks or activities for which your team uses VR?"
2. **Team Communication and Collaboration**:
    - "How has VR affected the way you communicate with your team during remote work?"
    - "Do you feel that VR helps or hinders collaboration compared to traditional video conferencing tools?"
3. **Team Dynamics and Relationships**:
    - "Has the use of VR influenced your sense of connection with colleagues?"
    - "Do you think VR has changed the way your team members interact or build relationships?"
4. **Productivity and Efficiency**:
    - "Do you feel more or less productive when using VR for meetings or collaborative tasks?"
    - "Has VR introduced any challenges or improvements in task management during remote projects?"

5. **Emotional and Psychological Effects**:
    - "How does using VR for work affect your mental well-being or stress levels?"
    - "Have you experienced any form of digital fatigue or VR-related exhaustion?"
6. **Adaptation and Learning Curve**:
    - "What was the learning curve like for adapting to VR tools in the workplace?"
    - "Has your company provided sufficient support for transitioning to VR?"
7. **Future of VR in Remote Work**:
    - "Do you see VR playing a larger role in the future of remote work? Why or why not?"
    - "What changes or improvements would you suggest for using VR in team collaboration?"
8. **Closing Questions**:
    - "What would you like to share about your experience using VR that we haven't covered?"
    - "Is there anything else you think is important regarding the impact of VR on team dynamics?"

This structure balances practical questions with exploratory ones, allowing the participants to share insights on their personal experiences while ensuring the interview touches on the main themes relevant to remote work and VR.

PREDEFINED THEMES FOR DEDUCTIVE ANALYSIS

For deductive thematic analysis, we will use predefined themes based on existing literature around team collaboration, technology adoption, and the psychological effects of immersive environments like VR.

Predefined Themes:

1. **Communication and Interaction**:
   - **Focus**: How VR impacts communication within teams and between team members.
   - **Sub-themes**: Clarity of communication, ease of collaboration, informal vs. formal interactions, non-verbal cues in VR.
2. **Team Cohesion and Relationships**:
   - **Focus**: The effect of VR on team dynamics and interpersonal relationships.
   - **Sub-themes**: Team bonding, social connection, trust-building, remote team integration.
3. **Productivity and Workflow**:
   - **Focus**: How VR affects individual and team productivity in remote work settings.
   - **Sub-themes**: Task efficiency, focus, workflow management, multitasking in VR.
4. **Emotional and Psychological Impact**:
   - **Focus**: The mental and emotional effects of using VR as a work tool.
   - **Sub-themes**: Stress, digital fatigue, cognitive load, immersive overload, feelings of isolation or presence.
5. **Adaptation and Learning Curve**:
   - **Focus**: The ease or difficulty of adopting and adapting to VR technologies in the workplace.
   - **Sub-themes**: Training, learning challenges, skill acquisition, company support.
6. **Perception of Future Use**:
   - **Focus**: Employee perceptions of the long-term viability and future of VR in remote work.
   - **Sub-themes**: Scalability, long-term adaptation, potential for innovation, barriers to adoption.

These predefined themes will guide our deductive analysis, with ChatGPT assisting in identifying relevant codes from the interview data. The themes offer a structured way to analyze the impact of VR on remote work while leaving room for insights specific to this emerging technology.

Hypothetical Interview Transcript

Interviewer: How long have you been using VR tools in your remote work environment, and what kinds of tasks or activities does your team typically use VR for?

Participant: I have been using VR for roughly eight months now. Initially, we only used it for occasional meetings—usually when we needed to brainstorm or have more interactive discussions. But over time, it has become more of a regular thing. Now, we use it for bigger team meetings, virtual training sessions, and creative workshops. It is particularly useful when we need to collaborate visually, like when we are sketching ideas on a virtual whiteboard or reviewing 3D models. I would say it is more hands-on compared to a regular video call, but we do not use it for everything. Sometimes, it is still easier just to hop on Zoom.

Interviewer: How has VR affected the way you communicate with your team during remote work?

Participant: It is definitely changed the way we communicate, and for the most part, it has been an improvement. In VR, it feels like we are all in the same room, which adds a level of engagement you do not get with traditional video calls. The avatars give a sense of presence, and even though it is virtual, it helps with things like body language and non-verbal cues. I find that I am more focused during VR meetings—it is harder to get distracted when you are fully immersed.

That being said, there are some challenges. Not everyone on the team is comfortable using VR, so there is a bit of a gap in participation. Some people are still hesitant, and that can slow things down. Plus, the tech learning curve at the start was steep—there were issues with headsets not working properly, or people just not being familiar with how to navigate the virtual space. It has been a bit smoother now, but there is still the occasional glitch or someone getting lost in the virtual room.

Interviewer: Has using VR influenced your sense of connection with your colleagues? Do you think it is changed the way you interact with team members?

Participant: In some ways, it is definitely brought us closer. The virtual environment makes it feel like we are in the same space, even if we are all working from different locations. I have noticed that when we use VR for social events or team-building exercises, it feels more interactive than just doing a Zoom happy hour or something. For example, we did a virtual coffee break in VR, and it felt more natural to move around and talk to different people, like you would in a real office setting.

But it is not perfect. There is still something missing. The avatars are fun, but they are not real people, you know? I still miss seeing facial expressions and real body language. Sometimes, VR feels a bit artificial, and I think that creates a little bit of emotional distance. It is definitely better than just staring at a screen of faces on Zoom, but it is not a replacement for face-to-face interaction.

Interviewer: How has VR impacted your productivity? Do you feel more or less productive when using VR for meetings or collaborative tasks?

Participant: That really depends on the task. For things like creative brainstorming or collaborating on designs, VR is amazing. It lets us visualize ideas and work on things together in a way that feels very collaborative. For those kinds of tasks, I would say I am definitely more productive. The fact that you can work on a virtual whiteboard or interact with 3D models speeds up the process and makes everything more hands-on.

But for everyday meetings? Not so much. It is kind of overkill for routine stuff like status updates or quick check-ins. There is also the setup time—you need to put on the headset, make sure everything is working, and sometimes deal with technical hiccups. Plus, meetings tend to run longer in VR because we are still figuring out the best way to use it. So, while it is great for certain tasks, I wouldn't say it's boosted my productivity across the board.

Interviewer: How has VR affected your mental and emotional well-being?

Participant: I did not expect VR to be as tiring as it is. It can be exhausting, especially if you are in it for long periods. There's so much sensory input in

VR—visuals, sounds, and the whole immersive environment—that it can get overwhelming after a while. I have found that after a couple of hours in VR, I feel drained in a way I do not after a regular video call. I have definitely experienced what I would call "VR fatigue," where I just need to take off the headset and take a break.

Another thing is the disconnect between being in this immersive virtual space and then taking off the headset and realizing you are just alone in your room. It can feel a bit jarring, and that is something I had not anticipated. It is like your brain is in two places at once, and it can take a little while to adjust back to reality after a long VR session.

Interviewer: What was the learning curve like for adapting to VR tools in the workplace? Has your company provided enough support for the transition to VR?

Participant: The learning curve was pretty steep at the beginning. Getting used to the controls, navigating the virtual environment, and even just wearing the headset for extended periods took some time. My company did provide training, but it was mostly self-guided, which worked for some of us but was a bit harder for others. The more tech-savvy people adapted quickly, but I know some colleagues struggled with it, especially at the start. We could have used more structured training or one-on-one sessions to help everyone get up to speed.

On the plus side, the company has been pretty good about providing the equipment and troubleshooting when things go wrong. They have been understanding that VR is new for most of us, and they've been patient as we all get used to it. But I do think more hands-on support early on would have made the transition smoother.

Interviewer: Do you think VR will play a bigger role in remote work in the future? Why or why not?

Participant: I think VR has a lot of potential for remote work, especially for certain industries and tasks. It is great for creative collaboration, training, and any situation where you need to visualize things in 3D. But I do not see it replacing everything. For routine meetings or quick conversations, regular video calls are still easier and more efficient.

That said, I do think as the technology improves, we will see more people adopting it. Right now, the hardware is still a bit bulky, and the software can be glitchy, but as those issues get ironed out, I think VR could become a bigger part of how we work remotely. But for now, I see it as just one tool in the toolbox—not the main way we do things.

Interviewer: What would you change about how VR is used in your workplace?

Participant: I think we need clearer guidelines on when to use VR and when it is better to stick with traditional tools. Right now, it feels like we are still experimenting, and sometimes we use VR for meetings where it is not really necessary. I would also suggest keeping VR meetings shorter—after an hour or so, it gets pretty tiring. And, as I mentioned earlier, more structured training would have helped, especially for the people who were not familiar with the technology.

Interviewer: Is there anything else you would like to share about your experience using VR for remote work?

Participant: Overall, I think VR is an exciting tool with a lot of potential. It is changed how I think about remote work and made certain tasks a lot more interactive. But it is not a magic bullet, and there are definitely challenges—especially around fatigue and learning to use the tech. I think with the right balance and support, VR could be a game-changer for certain aspects of remote work, but it is not going to replace everything.

SAFEGUARDING PRIVACY IN AI-ASSISTED RESEARCH

Ensuring data privacy and compliance is a critical aspect of using AI like ChatGPT for qualitative research. By anonymizing data, obtaining informed consent, adhering to data protection laws, securing storage, understanding OpenAI's policies, and maintaining transparency in reporting, you protect your participants and maintain the integrity of your research.

In the next section, we will move on to **Effective Prompt Engineering**, where we will explore how to create specific prompts that guide ChatGPT in providing accurate, insightful responses aligned with your research objectives.

Chapter Key Points:

- *Researchers can use default GPT models or build custom GPTs to employ ChatGPT in qualitative data analysis.*
- *Default GPT models are easily accessible and require minimal technical skills, suitable for general qualitative tasks.*
- *Custom GPTs allow for tailored analysis by providing specific instructions without the need for programming or APIs.*
- *Proper data preparation—including cleaning and formatting—is essential for accurate AI-assisted qualitative analysis.*
- *Breaking down large datasets into smaller chunks accommodates ChatGPT's token limits and maintains analysis context.*
- *Ensuring data privacy and compliance involves anonymizing data, obtaining informed consent, and adhering to data protection laws.*
- *Researchers must implement data security measures like encryption and access controls to protect sensitive data during AI analysis.*
- *Transparency and ethical reporting require documenting AI's role, acknowledging its limitations, and ensuring human judgment guides research findings.*

Chapter 10.

STEP-BY-STEP QUALITATIVE DATA ANALYSIS WITH CHATGPT AND QUALCODER

This section provides a comprehensive, easy-to-follow, step-by-step guide for conducting AI-assisted open coding using ChatGPT. Whether you are adopting a deductive or inductive approach, this guide will walk you through cleaning and preparing the interview transcript, setting up ChatGPT, creating effective prompts, and integrating the results with qualitative analysis software like QualCoder.

This chapter is essential for ensuring your analysis is thorough, transparent, and valid, leveraging the power of AI while maintaining researcher-driven insights.

STEP 1: CLEAN THE INTERVIEW TRANSCRIPT

Objective:

Before using ChatGPT for qualitative data analysis, ensure your data is clean, anonymized, and properly formatted (Figure 10.1).

Process:

1. **Remove Unnecessary Text**:
   - Go through the interview transcripts and remove any extraneous information that might interfere with the analysis. This includes:
     - **Filler Words**: Words like "um," "uh," and other verbal fillers.
     - **Non-meaningful Phrases**: Phrases that do not contribute to the research questions, such as unrelated comments or interruptions.
     - **Time Stamps/Interviewer Prompts**: If transcripts include time stamps, interviewer questions, or prompts that are not needed for the analysis, remove or minimize them.
2. **Anonymize Sensitive Information**:
   - To protect participants' privacy, remove or replace any personal identifiers, such as names, company information, or geographical details.
   - Replace sensitive details with generic labels:
     - Example: Replace "John Smith from ABC Corp." with "Participant 1 from Company A."
   - If you are using multiple interviews, ensure that each participant has a consistent identifier (e.g., "Participant 1") throughout the dataset for clarity and anonymity.
3. **Ensure Data is Structured for AI**:
   - Clearly distinguish between interviewer and participant. For instance:
     - **Interviewer**: "How do you feel about using VR for team meetings?"
     - **Participant**: "I think it's both exciting and challenging."

This step ensures that ChatGPT focuses on the relevant participant responses for analysis as you can see in the hypothetical interview example in the preceding sections in Chapter 4.

Figure 10.1. Cleaning the Interview Transcript

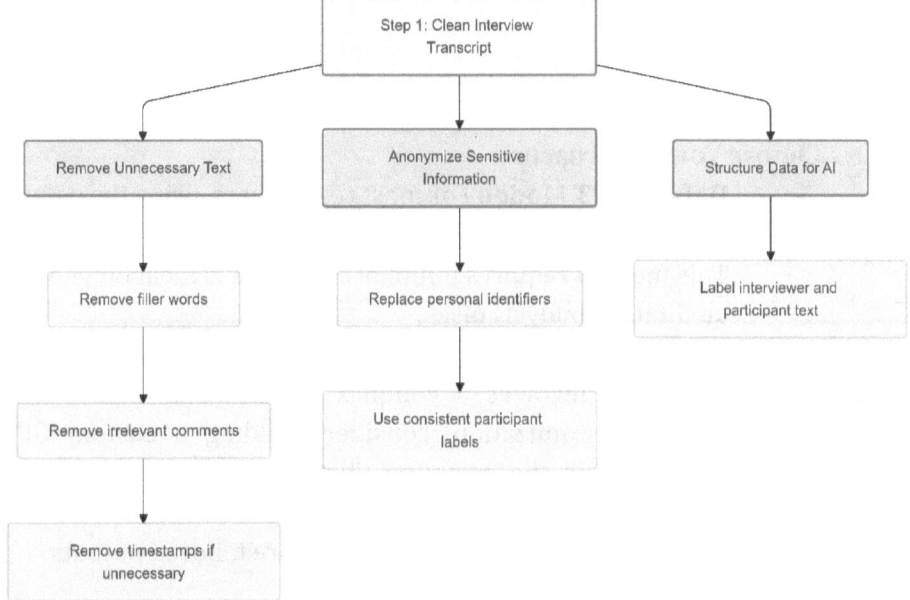

STEP 2: SET UP CHATGPT FOR QUALITATIVE DATA ANALYSIS

Objective:

Configure ChatGPT for effective use in qualitative data analysis, either by using the default GPT models or by building a custom GPT tailored to your study (Figure 10.2).

Process:

1. **Choose Your Approach:**
   - **Default GPT Model:** For most researchers, using the default GPT-4 model through the OpenAI web interface will suffice. This method requires minimal setup and is ideal for general qualitative analysis tasks.
   - **Custom GPT:** If your research requires highly specific coding or involves a complex dataset that necessitates deeper customization, consider building a custom GPT model within the premium ChatGPT version. This option allows you to personalize the model with your specific dataset and prompts, resulting in more tailored outputs.
2. **Configure the Model:**
   - For both approaches, ensure you configure ChatGPT with specific instructions regarding your qualitative research context. When using a custom GPT, provide the model with relevant background information, coding frameworks, or previously coded data to fine-tune it for your specific needs.
3. **Manage Token Limits:**
   - Be mindful of the token limits (i.e., the number of words or characters ChatGPT can process in a single query). If necessary, split larger transcripts into smaller chunks for efficient processing.

Note: While custom GPTs can be built through Application Programming Interface (API) integrations for highly specific requirements, this book focuses exclusively on utilizing ChatGPT's default and premium custom GPT features available directly through the ChatGPT platform. The process of building and deploying custom GPTs via APIs is therefore outside the scope of this book, allowing us to concentrate on accessible, user-friendly applications within the ChatGPT interface.

Figure 10.2. Setting Up ChatGPT for Qualitative Data Analysis

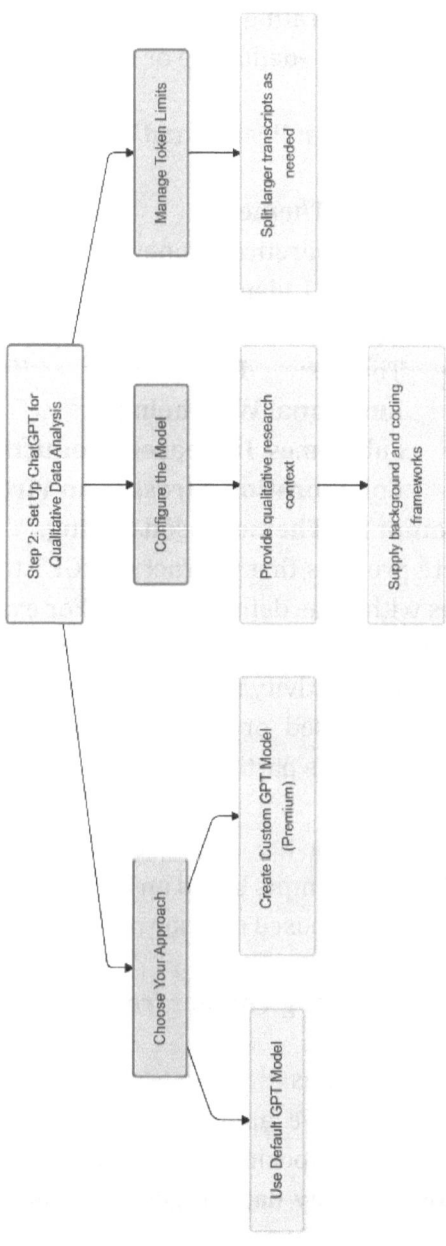

Step 3: Define Themes and Develop Prompts for Deductive or Inductive Analysis

Objective:

Develop a strategy for generating initial codes based on your chosen qualitative analysis approach—deductive or inductive (Figure 10.3).

Deductive Thematic Analysis: Predefined Themes and Subthemes

1. **Identify Predefined Themes**:
   - Review the theoretical framework or literature that informs your study and identify the **key themes** and **subthemes** relevant to your research. For example:
     - Themes: **Productivity**, **Team Dynamics**, **Emotional Well-being**.
     - Subthemes: **Increased Productivity**, **Challenges in Collaboration**, **Stress from Virtual Tools**.
2. **Develop Prompt for Theme-Based Coding**:
   - Create prompts that instruct ChatGPT to search for specific codes within the defined themes. For example:
     - **Prompt**: "Identify any statements related to productivity in this transcript. Focus on mentions of increased or decreased productivity, and explain how the participant describes these changes."
3. **Refinement**:
   - After receiving the initial codes from ChatGPT, refine or modify the prompts based on the response quality, ensuring the AI stays focused on your predefined themes.

Inductive Analysis: Generating Themes from Data

1. **No Predefined Themes**:
   - In the inductive approach, let the data speak for itself. Instead of imposing predefined themes, ChatGPT will explore the raw data to generate **codes** and **themes** that emerge naturally.
2. **Develop Open-Ended Prompts**:

- Use prompts that allow for open-ended exploration of the data. For instance:
    - **Prompt**: "Analyze this transcript and identify any recurring ideas or patterns. Focus on concepts related to team collaboration and emotional responses to using VR."

3. **Iterative Process**:
    - Review ChatGPT's initial output and adjust your prompts to focus on emerging ideas that seem most relevant. You may need to run multiple iterations to explore different angles within the data.

Figure 10.3. Defining Themes and Developing Prompts for Analysis

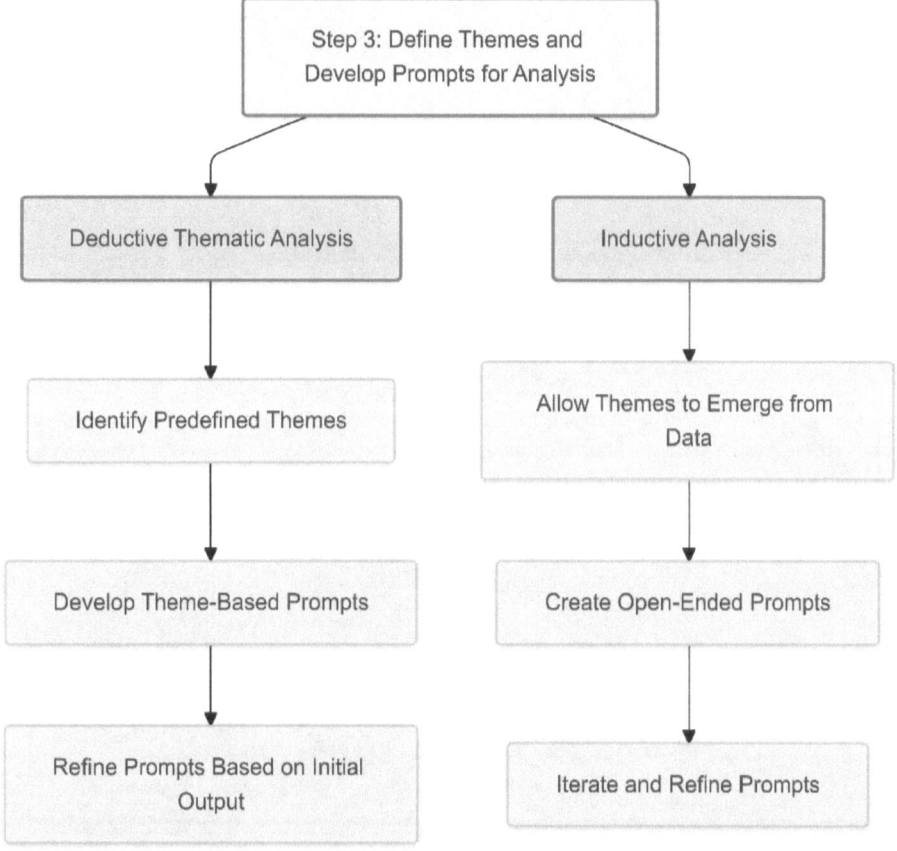

The following sections illustrate how the ChatGPT prompt identifies relevant codes from each section of the interview transcript, guided by predefined thematic areas and structured through deductive thematic analysis.

ChatGPT prompt Example:

Instructions:

You are tasked with performing a deductive thematic analysis on an interview transcript using a predefined thematic framework. Your goal is to identify relevant codes within the transcript that correspond to the provided themes and sub-themes. For each identified code, extract the exact quotes from the transcript that support it.

Please follow these steps:

1. **Read the Interview Transcript Carefully:**
   - Thoroughly read the entire transcript to understand the context and content.
2. **Use the Predefined Thematic Framework:**
   - Refer to the provided themes and sub-themes as a guide for your analysis.
3. **Identify Relevant Codes:**
   - For each theme and sub-theme, identify codes that are evident in the transcript.
   - Codes should be concise phrases that capture key ideas or concepts related to the themes.
4. **Extract Exact Quotes:**
   - For every identified code, find and extract the exact quote(s) from the transcript that illustrate it.
   - Ensure that the quotes are accurate and include any relevant context.
5. **Organize Your Findings:**
   - Present your results in a structured format, organized by theme and sub-theme.
   - Under each sub-theme, list the identified codes followed by their supporting quotes.
6. **Ensure Accuracy and Relevance:**
   - Make sure all codes and quotes directly relate to the themes and sub-themes.
   - Avoid introducing new themes not included in the predefined framework.

The following is the predefined thematic framework for your reference.

Predefined Thematic Framework:

1. **Communication and Interaction:**
   - **Focus:** How VR impacts communication within teams and between team members.
   - **Sub-themes:**
     - Clarity of Communication
     - Ease of Collaboration
     - Informal vs. Formal Interactions
     - Non-Verbal Cues in VR
2. **Team Cohesion and Relationships:**
   - **Focus:** The effect of VR on team dynamics and interpersonal relationships.
   - **Sub-themes:**
     - Team Bonding
     - Social Connection
     - Trust-Building
     - Remote Team Integration
3. **Productivity and Workflow:**
   - **Focus:** How VR affects individual and team productivity in remote work settings.
   - **Sub-themes:**
     - Task Efficiency
     - Focus
     - Workflow Management
     - Multitasking in VR
4. **Emotional and Psychological Impact:**
   - **Focus:** The mental and emotional effects of using VR as a work tool.
   - **Sub-themes:**
     - Stress
     - Digital Fatigue
     - Cognitive Load
     - Immersive Overload
     - Feelings of Isolation or Presence
5. **Adaptation and Learning Curve:**

- o **Focus:** The ease or difficulty of adopting and adapting to VR technologies in the workplace.
- o **Sub-themes:**
  - Training
  - Learning Challenges
  - Skill Acquisition
  - Company Support

6. **Perception of Future Use:**
   - o **Focus:** Employee perceptions of the long-term viability and future of VR in remote work.
   - o **Sub-themes:**
     - Scalability
     - Long-Term Adaptation
     - Potential for Innovation
     - Barriers to Adoption

Interview Transcript:

[Please insert the interview transcript below this line.]

Example of How to Present Your Findings:

Theme 1: Communication and Interaction

- **Sub-theme:** Clarity of Communication
  - **Code:** Improved understanding through immersive meetings
    - **Quote:** "Since we started using VR, I feel like I understand my colleagues' ideas much better because we can visualize concepts together."
- **Sub-theme:** Non-Verbal Cues in VR
  - **Code:** Lack of real facial expressions
    - **Quote:** "It's sometimes hard to gauge reactions because avatars don't show real facial expressions."

Repeat this format for each theme and sub-theme.

Note:

- *Focus on quality over quantity; it is better to have well-supported codes than a large number of weak ones.*
- *Ensure that the extracted quotes are directly related to the codes and provide clear evidence of the themes.*

STEP 4: CREATE PROMPTS AND RUN THE CODING PROCESS

Objective:

Execute the coding process by running tailored prompts for each transcript and saving the outputs for further analysis (Figure 10.4).

Process:

1. **Create Clear and Targeted Prompts**:
   - For each transcript, create prompts that are aligned with your research objectives. Example prompts might include:
     - **Deductive Example**: "For this transcript, identify any references to team communication. Focus on how participants describe their interactions and challenges when using VR."
     - **Inductive Example**: "Identify any emerging themes related to how participants experience emotional well-being during virtual meetings. Summarize key points mentioned."
2. **Run the Prompts for Each Transcript**:
   - Submit each transcript to ChatGPT using the prepared prompts. Ensure that the AI's output is saved in a structured format (such as a Word document or text file) after each run.
   - **Example Output**:
     - Codes: **Team Isolation, Increased Focus in VR Meetings, Stress from Learning New Tools**.
     - Quotes: "I find that working in VR makes me feel disconnected from my team sometimes."
3. **Iterate and Validate**:
   - For each transcript, you may need to run multiple prompts to explore different aspects of the data (e.g., team dynamics, emotional responses). Use **iterative questioning** to refine the output. For example, if a code seems too broad, ask ChatGPT to break it down into more specific components:
     - **Prompt**: "Can you provide more detail on the different ways participants describe 'team isolation'? What specific factors contribute to this?"
4. **Save Outputs for Analysis**:

- After running each transcript, ensure the outputs are saved in a consistent format. These outputs will later be uploaded into QualCoder for further analysis.

Figure 10.4. Creating Prompts and Running the Coding Process

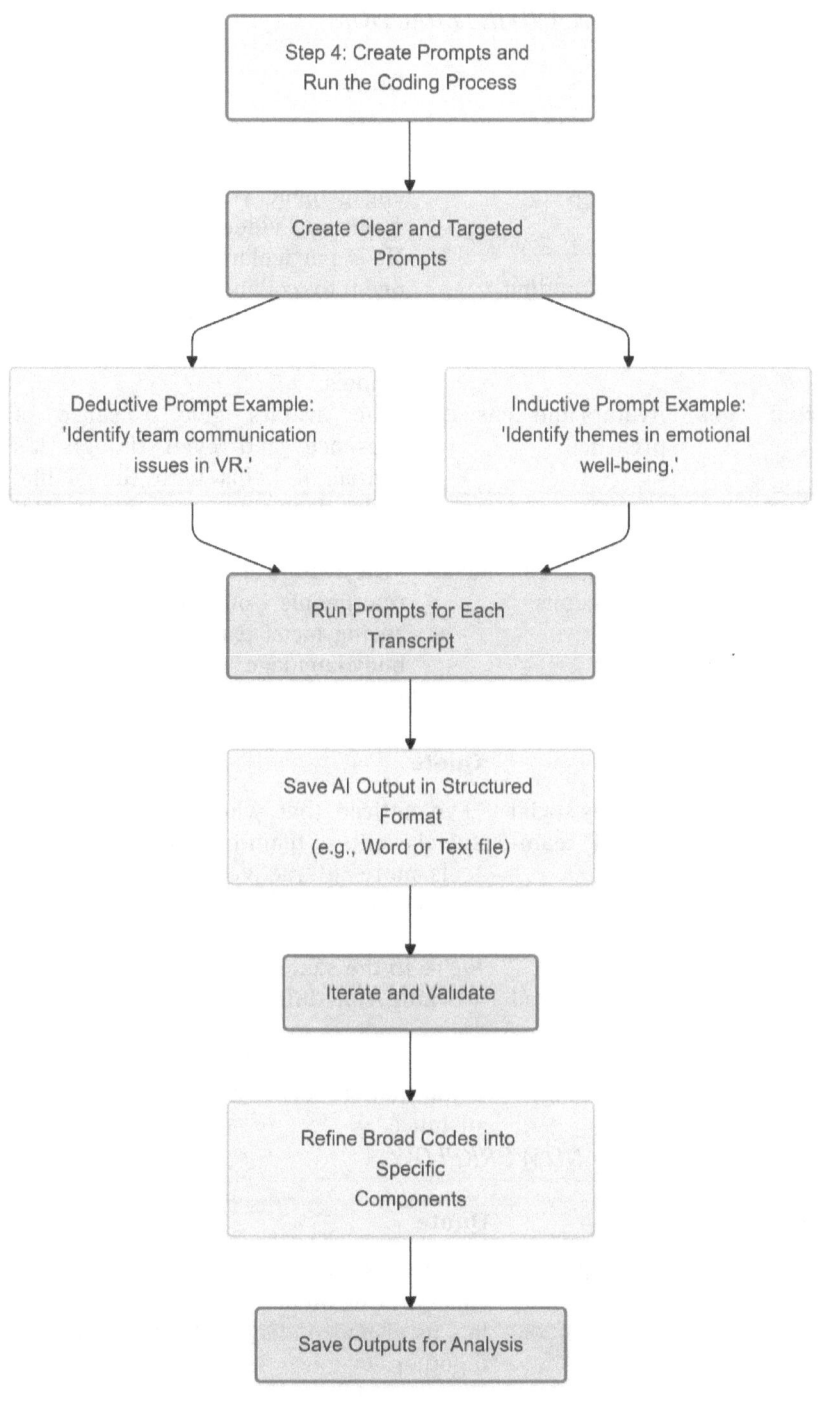

Output from ChatGPT

THEME 1: COMMUNICATION AND INTERACTION

Sub-theme	Code	Quote
Clarity of communication	Enhanced engagement in VR meetings	"In VR, it feels like we are all in the same room, which adds a level of engagement you don't get with traditional video calls."
	Better understanding through visual collaboration	"It is particularly useful when we need to collaborate visually, like when we are sketching ideas on a virtual whiteboard or reviewing 3D models."
Non-verbal cues in vr	Avatars and sense of presence	"The avatars give a sense of presence, and even though it's virtual, it helps with things like body language and non-verbal cues."
	Missing real facial expressions	"The avatars are fun, but they're not real people, you know? I still miss seeing facial expressions and real body language."

THEME 2: TEAM COHESION AND RELATIONSHIPS

Sub-theme	Code	Quote
Team bonding	VR enhances social events and team-building	"I've noticed that when we use VR for social events or team-building exercises, it feels more interactive than just doing a Zoom happy hour or something."
Social connection	Increased connection through shared virtual space	"The virtual environment makes it feel like we're in the same space, even if we're all working from different locations."
	Emotional distance in virtual settings	"Sometimes, VR feels a bit artificial, and I think that creates a little bit of emotional distance."

THEME 3: PRODUCTIVITY AND WORKFLOW

Sub-theme	Code	Quote
Task efficiency	VR supports creative and visual tasks	"For things like creative brainstorming or collaborating on designs, VR is amazing. It lets us visualize ideas and work on things together."
Workflow management	VR is less suitable for routine tasks	"For everyday meetings? Not so much. It is kind of overkill for routine stuff like status updates or quick check-ins."

Multitasking in vr	Longer setup time for VR meetings	"There's also the setup time—you need to put on the headset, make sure everything is working, and sometimes deal with technical hiccups."

THEME 4: EMOTIONAL AND PSYCHOLOGICAL IMPACT

Sub-theme	Code	Quote
Digital fatigue	Experience of VR-induced fatigue	"I didn't expect VR to be as tiring as it is. It can be exhausting, especially if you're in it for long periods."
Immersive overload	Overwhelming sensory input in VR	"There's so much sensory input in VR—visuals, sounds, and the whole immersive environment—that it can get overwhelming after a while."
Feelings of isolation or presence	Adjustment challenges post-VR sessions	"Another thing is the disconnect between being in this immersive virtual space and then taking off the headset and realizing you're just alone in your room."

THEME 5: ADAPTATION AND LEARNING CURVE

Sub-theme	Code	Quote
Learning challenges	Steep learning curve for VR tools	"The learning curve was pretty steep at the beginning. Getting used to the controls, navigating the virtual environment... took some time."
Company support	Mixed experiences with training support	"My company did provide training, but it was mostly self-guided, which worked for some of us but was a bit harder for others."

THEME 6: PERCEPTION OF FUTURE USE

Sub-theme	Code	Quote
Scalability	Limited potential for routine tasks	"For routine meetings or quick conversations, regular video calls are still easier and more efficient."
Potential for innovation	Growing adoption as technology improves	"As the technology improves, I think we'll see more people adopting it... VR could become a bigger part of how we work remotely."

STEP 5: IMPORT DATA INTO QUALCODER AND GENERATE A CODE TREE

QUALITATIVE DATA CODING: MANUAL VS. AUTOMATED IMPORT METHODS IN QUALCODER

In qualitative analysis, importing and organizing data within QualCoder can be approached in three distinct ways: **Manual Coding**, **Semi-Automated Import with ChatGPT Assistance**, and **Fully Automated Import via API Integration**. Each method has unique benefits and limitations, and the best approach depends on factors such as research needs, available time, technical skills, and desired engagement with the data (Figure 10.5).

Only the **Manual Coding** option (Option 1) falls within the scope of this book, covering the complete process of importing, coding, and organizing data manually within QualCoder. The **Semi-Automated Import** and **Fully Automated Import via API Integration** (Options 2 and 3) are outside the scope of this book and require advanced technical knowledge and tools that go beyond this guide's focus.

1. MANUAL IMPORT AND CODING

Steps:

- **Step 1: Upload Transcript Files**
    - Navigate to **File > Import** in QualCoder.
    - Select and upload qualitative data files, such as interview or focus group transcripts.
    - Organize and label the imported files by participant, interview number, or relevant metadata.
- **Step 2: Manual Coding Based on ChatGPT Insights**
    - Begin by reviewing ChatGPT's outputs, including suggested codes, themes, and potentially useful quotes or phrases.
    - In QualCoder, create new codes based on these suggestions, or modify them as necessary to suit your research context.
    - Open each transcript and manually review the text. Highlight relevant sections and assign codes as you proceed.
    - By manually applying codes, you control the interpretation of ChatGPT's insights and engage closely with the data.

Step 3: Build a Code Tree

- As you code, you can structure related codes into themes and sub-themes, forming a hierarchical code tree.
- Group similar codes, label overarching themes, and make adjustments based on your understanding of the data and ChatGPT's suggestions.
- This iterative process ensures that you remain connected to the data while benefiting from ChatGPT's automated suggestions.

Pros of Manual Coding:

- **Deep Engagement**: Allows you to engage fully with the data, improving the quality and validity of insights.
- **Complete Control**: Researchers retain control over which sections to code, ensuring that interpretations are researcher-led.
- **Flexibility**: You can modify ChatGPT's suggestions, adapting the codes and themes as new patterns or insights emerge during coding.

Cons of Manual Coding:

- **Time-Intensive**: Manually applying codes across lengthy transcripts can be time-consuming, especially for large datasets.
- **Risk of Inconsistency**: In manual coding, there may be variations in coding style or focus, depending on the researcher's interpretation at different stages.
- **Labor-Intensive Code Structuring**: Organizing codes into themes can require significant effort to maintain consistency and hierarchy.

2. AUTOMATED IMPORT AND CODING (WITH CHATGPT)

The automated method uses ChatGPT to generate codes and themes, which are then imported into QualCoder with minimal manual intervention. While currently limited in full automation, this method offers a semi-automated approach to streamline the coding process (Table 10.1).

Steps:

- **Step 1: Open Coding in ChatGPT**
    - Use ChatGPT to perform open coding by analyzing transcripts and suggesting codes, themes, and potential categories.
    - ChatGPT may provide direct quotes, phrases, or thematic insights that correspond to the interview data.
- **Step 2: Export ChatGPT's Output to a REFI-QDA or CSV Format**
    - Save ChatGPT's generated codes, themes, and quote suggestions in a CSV or REFI-QDA XML format.
    - The structure should include relevant details such as code, category, quote, and source.
- **Step 3: Import the Codes and Themes into QualCoder**
    - Navigate to **File > Import > REFI-QDA Project** or **Import > Codes from CSV** in QualCoder.
    - Select the file containing ChatGPT's codes and themes, and follow the prompts to import them.
    - QualCoder will import the codes and associate any provided quotes with relevant codes, allowing for partial automation of the coding process.
- **Step 4: Review and Refine Imported Codes**
    - Open each transcript in QualCoder and examine how well the automated codes match the data.
    - Adjust, delete, or reassign codes where necessary to ensure accuracy and alignment with your research questions.
    - If code application is inconsistent, manually apply or modify codes based on the context.

Pros of Automated Import:

- **Efficiency**: Reduces the time needed to create and apply codes manually, especially useful for large datasets.
- **Quick Initial Structure**: Provides an initial code structure that can be refined, giving researchers a head start in analysis.
- **Scalability**: Allows for easier analysis of large datasets by providing a foundational code set that can be manually refined.

Cons of Automated Import:

- **Reliance on AI Interpretation**: Automated codes are based on ChatGPT's interpretation, which may not capture nuanced meanings as a researcher would.
- **Potential for Data Loss**: Importing codes without careful review may lead to missing or misinterpreted insights.
- **Limited Control Over Contextualization**: ChatGPT may not accurately apply codes based on contextual factors, requiring additional manual adjustments in QualCoder.

Table 10.1. Manual vs. Automated import in QualCoder

Aspect	Manual import and coding	Automated import and coding (with chatgpt)
Engagement level	High	Moderate
Time requirement	High	Lower
Control over data	High	Moderate (depends on quality of ChatGPT coding)
Coding accuracy	Researcher-driven; more accurate context analysis	AI-driven; may require refinement
Scalability	Suitable for smaller datasets	Better suited for large datasets
Data consistency	Prone to researcher variation	Consistent but may lack nuanced interpretation

3. Fully Automated Import via API Integration

A third option—**Fully Automated Import via API Integration**—could be considered a viable solution, especially for large-scale qualitative projects or research teams requiring streamlined workflows (Table 10.2). Here is how it could be framed:

The most advanced method of integrating ChatGPT and QualCoder uses APIs to automate both analysis and data transfer. While QualCoder does not have a built-in API, creating a custom middleware that connects OpenAI's API for ChatGPT with QualCoder can make fully automated coding possible.

Steps:

- **API Setup:**
    - Obtain API access through OpenAI's platform and set up an environment where ChatGPT can receive and analyze transcripts directly.
    - Develop or utilize a middleware solution that can translate data from ChatGPT into a format QualCoder recognizes.
    - This might require coding knowledge, as QualCoder lacks an out-of-the-box API. However, a custom script can serve as a bridge between ChatGPT's output and QualCoder's import functions.
- **Automated Coding and Importing:**
    - With APIs in place, ChatGPT analyzes transcripts and generates codes automatically. These codes, along with relevant quotes and themes, can then be structured for QualCoder in real time.
    - Middleware would transfer ChatGPT's output directly into QualCoder, applying codes and themes to the transcripts as they are processed.
    - This setup can achieve full automation by mapping ChatGPT-generated codes directly onto specific text segments in QualCoder without any manual intervention.
- **Efficient Data Management:**
    - This fully automated workflow is ideal for large datasets or fast-paced projects. By automating the full cycle—from text

analysis and coding by ChatGPT to seamless code application in QualCoder—this approach minimizes manual tasks and boosts efficiency.
- With all coding done in real-time, researchers can focus more on interpreting results rather than on data processing.

Pros of Fully Automated Import via API:

- **Maximal Efficiency**: Ideal for managing and coding large datasets quickly.
- **Seamless Workflow**: Provides a streamlined, end-to-end solution that automates qualitative data analysis, from initial coding to importing into QualCoder.
- **Scalability**: Particularly useful for projects requiring rapid or high-volume data processing.

Cons of Fully Automated Import via API:

- **Technical Setup**: Requires substantial programming knowledge to create a middleware or custom script for integration, which may be beyond the reach of some researchers.
- **Dependency on AI Interpretation**: Automated codes are based on ChatGPT's analysis, which may lack contextual nuance.
- **Potential Data Loss**: Like all automated systems, this setup could miss subtle insights or nuances, requiring some manual review.

Table 10.2. Fully Automated Import via API vs. Manual and Semi-Automated Approaches

Aspect	Manual coding	Semi-automated import	Fully automated import via api
Engagement level	High	Moderate	Low
Technical knowledge	Minimal	Moderate (basic data import/export)	High (programming skills required)
Coding accuracy	High (researcher-controlled)	Moderate (review of imported codes)	Variable (ChatGPT-driven coding)
Time requirement	High	Lower	Lowest
Best use case	Smaller datasets, detailed analysis	Medium datasets, efficient structuring	Large datasets, rapid processing

In this context, a fully automated API approach can offer a powerful solution for researchers seeking to integrate AI-driven coding directly with QualCoder. While it requires a substantial initial setup, it can be a transformative method for scaling up qualitative research projects. Including all three methods provides a comprehensive view of how different approaches to integration can meet various research needs, from deep manual engagement to high-efficiency automation.

Figure 10.5. Importing Data to QualCoder and Generating Code Tree

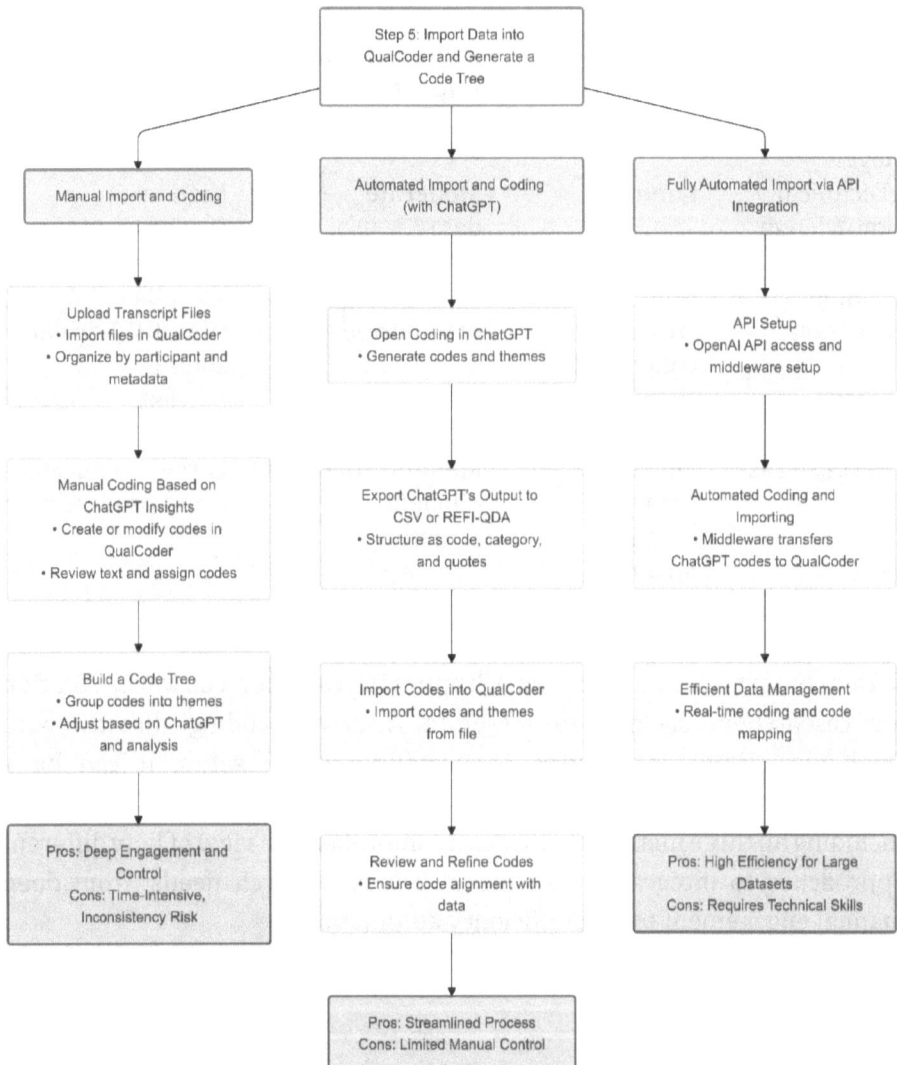

Choosing the Right Approach

- **The manual method** is ideal for those who prefer hands-on control and a deeper connection with the data. It is suited for researchers who want full responsibility over how codes are applied and refined.
- **The semi-automated approach** balances automation and manual oversight. It reduces repetitive tasks like code creation but still allows the researcher to manually apply codes to specific text segments, maintaining flexibility and interpretive depth.
- **The fully automated approach** is recommended for advanced users familiar with programming and API integration. It offers maximum efficiency, making it ideal for large-scale projects where speed and automation are crucial.

STEP 6: ANALYZE CODES IN QUALCODER

Objective:

Finalize the coding process and analyze themes across the entire dataset (Figure 10.6).

Process:

1. **Conduct Cross-Analysis**:
   - Using QualCoder, conduct cross-analysis to explore how themes vary across different participants or transcripts. For example, you could compare how different team members experience "isolation" when using VR for remote meetings.
2. **Generate Reports and Visualizations**:
   - QualCoder allows you to generate reports that summarize the frequency and distribution of codes across the dataset. Use these reports to explore relationships between themes and to visualize the results of your analysis.
3. **Interpret Findings**:
   - With the help of the code tree and detailed qualitative coding, interpret the findings to answer your research questions. Use the themes and quotes identified in ChatGPT and refined in QualCoder to draw meaningful conclusions from the data.

Figure 10.6. Analyzing Codes in QualCoder

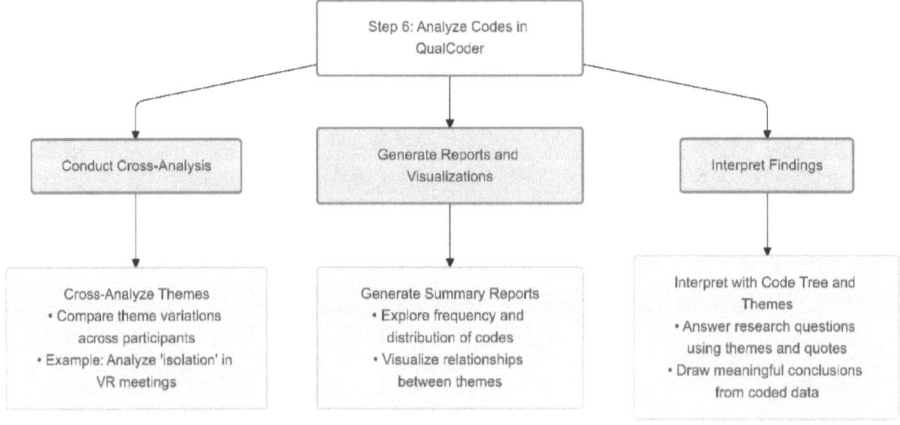

Step 7: Develop Themes and Categories

After the open coding process, move toward **thematic analysis** to identify larger patterns and connections within your data (Figure 10.7).

1. **Group Codes into Themes**:
   - Use QualCoder to group related codes into broader themes. For example, codes related to "communication challenges" and "technology adaptation" might form a theme like "Barriers to Remote Work Efficiency."
2. **Use ChatGPT for Thematic Suggestions**:
   - Prompt ChatGPT to assist in grouping codes into potential themes:
     Example Prompt: "Here are the codes I generated from my data. Can you suggest overarching themes or categories?"
   - This step allows you to leverage AI for deeper pattern recognition.
3. **Finalize Themes in QualCoder**:
   - Finalize the themes within QualCoder by reviewing both AI-suggested and manually created themes.
   - Refine as necessary, merging or adjusting themes as you see fit.

Figure 10.7. Developing Themes and Categories

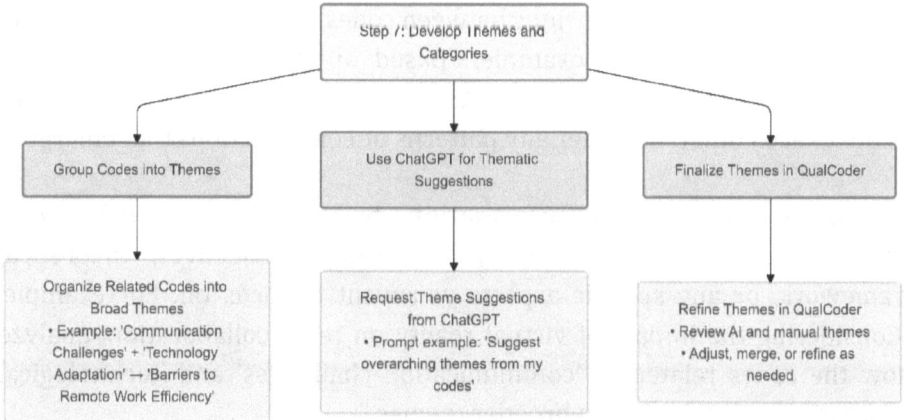

STEP 8: THEMATIC ANALYSIS IN CHATGPT

Objective:

Leverage ChatGPT to perform a deeper thematic analysis by identifying patterns, relationships, and overarching themes within your coded data (Figure 10.8).

Process:

1. **Prepare Summarized Data for Analysis:**

Compile Codes and Excerpts by gathering all the codes and their associated excerpts from QualCoder. Organize them into a structured format, such as a table or a document, grouping similar codes together.

Create a Summary Document that includes:

> **Code Names:** The labels you have assigned to different codes.
> **Representative Quotes:** Key excerpts that exemplify each code.
> **Initial Observations:** Any notes or insights you have made during the coding process.

2. **Develop Advanced Prompts for Thematic Exploration:**

Craft Prompts to Identify Relationships by creating prompts that ask ChatGPT to explore relationships between codes, potential sub-themes, and overarching themes. For example, "Based on the following codes and excerpts, identify any overarching themes and explain how these codes relate to each other. Consider any patterns or contradictions that emerge."

Incorporate Contextual Information by providing ChatGPT with background information about your research objectives, theoretical framework, or any specific aspects you want to focus on. For example, "Considering the impact of virtual reality on team collaboration, analyze how the codes related to 'communication challenges' and 'technological adaptation' interact to form broader themes."

3. **Run Thematic Analysis Prompts in ChatGPT:**

Input Summarized Data and Prompts by paste the summarized codes, excerpts, and your crafted prompts into ChatGPT. Ensure that the input is clear and well-organized to help the AI process the information effectively.

Review AI-Generated Themes. ChatGPT will generate potential themes, highlight relationships between codes, and may identify nuances that were not immediately apparent. For example, "The codes 'team isolation' and 'lack of non-verbal cues' suggest a broader theme of 'Challenges in Remote Communication.' Additionally, 'increased focus' and 'flexibility' point towards 'Benefits of VR in Productivity.'"

4. **Iteratively Refine Themes with ChatGPT:**

Ask Follow-Up Questions. If the initial output is too broad or lacks depth, ask ChatGPT to elaborate or focus on specific aspects. For example, "Can you delve deeper into how 'team isolation' impacts 'emotional well-being,' and suggest any sub-themes that emerge from this relationship?"

Explore Contradictions and Tensions by encourage ChatGPT to identify any contradictions or tensions within the data. For example, "Identify any conflicting themes or paradoxes within the participant responses related to 'productivity' and 'stress levels.'"

5. **Document AI-Suggested Themes:**

Save the Outputs by recording the themes and insights generated by ChatGPT for each prompt. Organize them in a way that aligns with your existing code structure for easy comparison.

Note AI Contributions: by clearly mark which themes and insights were suggested by ChatGPT to maintain transparency in your analysis process.

6. **Integrate AI Insights with Manual Analysis:**

Compare with Manual Themes by place the AI-generated themes alongside those you've developed manually in QualCoder. Identify overlaps, differences, and new perspectives introduced by ChatGPT.

Synthesize Themes by combine the most relevant and insightful themes from both AI and manual analyses. Refine the thematic structure to reflect a comprehensive understanding of the data.

Figure 10.8. Thematic Analysis in ChatGPT

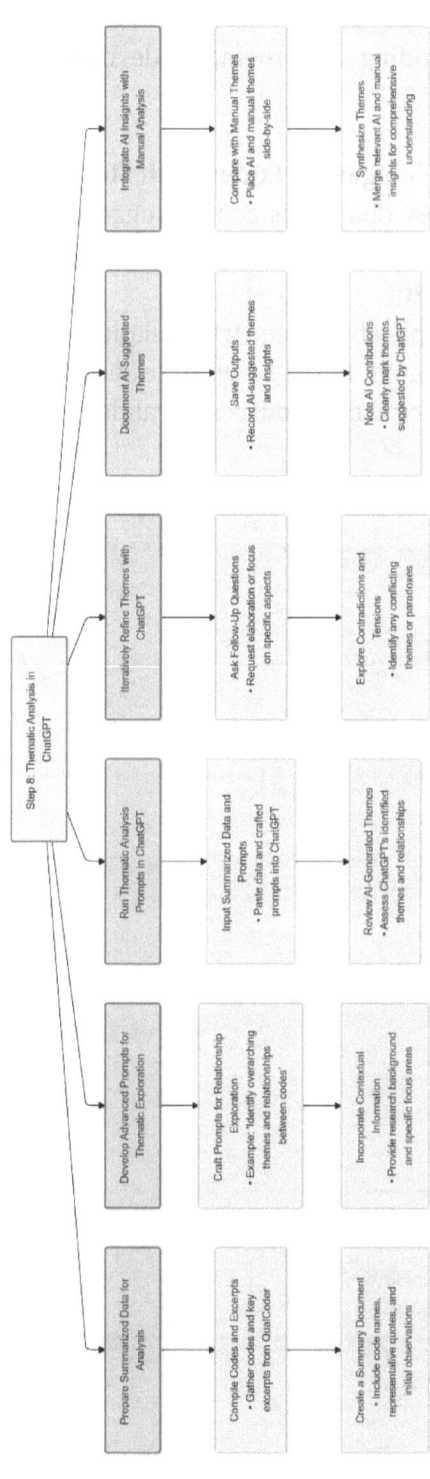

STEP 9: VALIDATE AND CROSS-CHECK

Ensure that your coding and thematic development are rigorous and transparent (Figure 10.9).

1. **Compare AI-Suggested Codes with Manual Coding**:
    - Review ChatGPT's coding suggestions against your own to ensure accuracy and to mitigate potential biases from the AI.
2. **Triangulation**:
    - If applicable, use triangulation by comparing the AI-generated findings with traditional manual methods or additional data sources to validate your results.

Figure 10.9. Validating and Cross-Checking

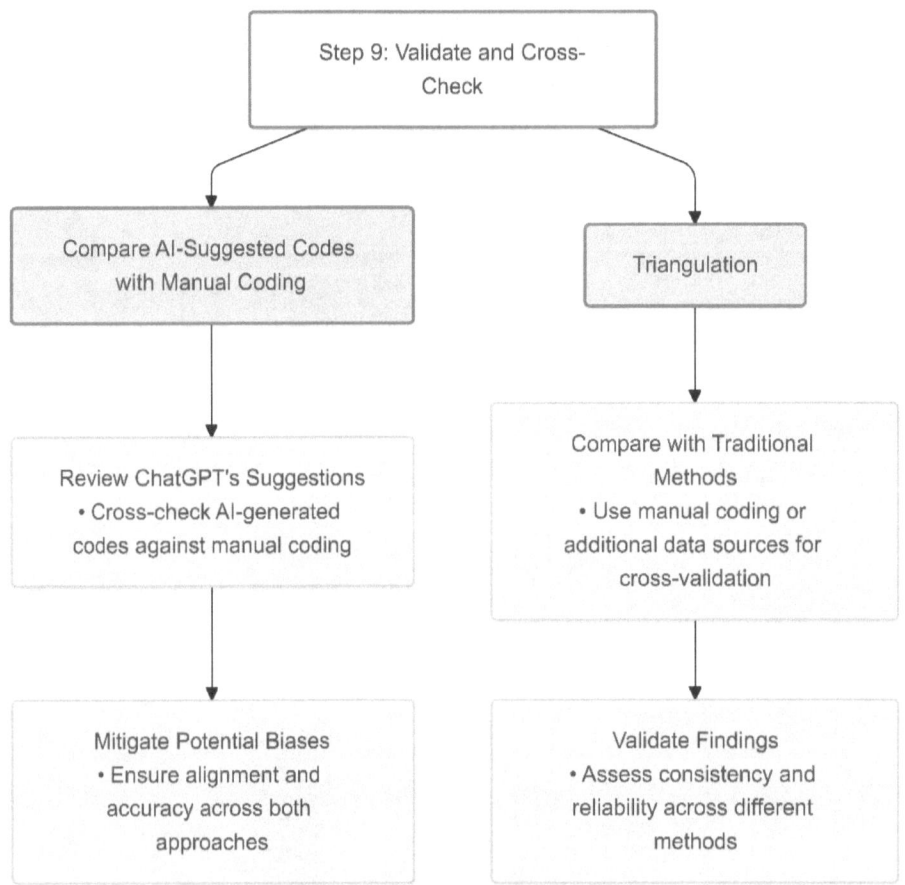

STEP 10: VISUALIZE AND ANALYZE IN QUALCODER

Use QualCoder's inbuilt tools for further analysis and visualization of your coded data (Figure 10.10).

1. **Generate Code Trees**:
   o Use QualCoder's **Code Tree** feature to visualize how your codes and themes are structured. This is useful for seeing the relationships between different themes and sub-themes.
2. **Memo Writing**:
   o Create memos within QualCoder to document your thoughts, interpretations, and reflections on the coding process. This helps in maintaining transparency and rigor.
3. **Export Findings**:
   o Once coding is complete, you can export the coded data, themes, and memos for further analysis or inclusion in your research report.

Figure 10.10. Visualizing and Analyzing in QualCoder

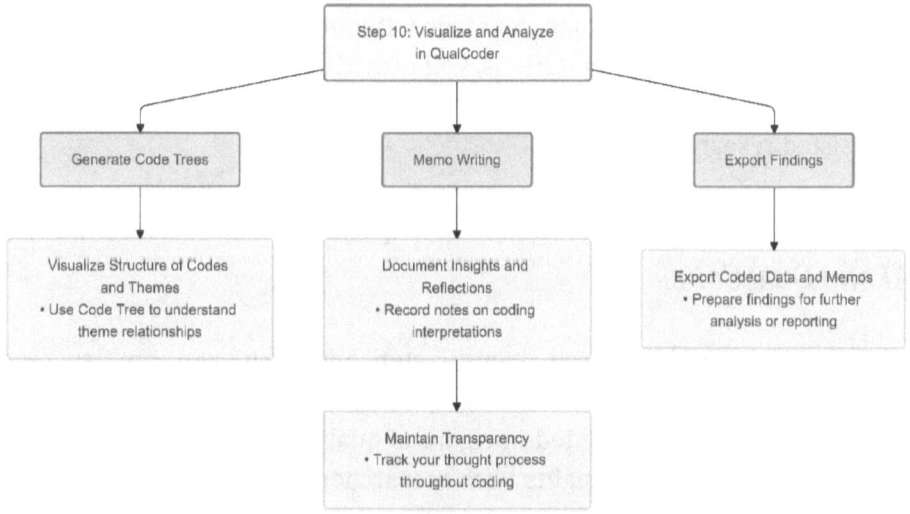

REPORT AND PRESENT FINDINGS

Present your findings, ensuring that the role of ChatGPT and QualCoder is clearly documented.

When presenting your findings, it is crucial to thoroughly explain how you integrated ChatGPT and QualCoder into your research process. Start by detailing their roles in your methodology, specifically how ChatGPT assisted with coding and thematic analysis. This transparency enhances the credibility of your study by allowing others to understand the extent of AI involvement and how it influenced your results.

Organize your findings around the themes you have developed in QualCoder. Use visual aids like code trees or thematic maps to illustrate these themes and their interconnections effectively. These visualizations can help readers grasp complex relationships within your data and add depth to your presentation.

It is also important to acknowledge any limitations encountered when using ChatGPT. Discuss challenges such as the AI's potential difficulty in fully capturing nuanced contextual meanings without human interpretation. By openly addressing these limitations, you provide a balanced perspective on your methodology and underscore the importance of human oversight in AI-assisted research.

ACHIEVING ROBUST AND TRANSPARENT QUALITATIVE ANALYSIS WITH AI AND QUALCODER

By following this structured, step-by-step approach, you can leverage ChatGPT to streamline the initial stages of open coding while ensuring that the analysis remains grounded in rigorous qualitative research principles. Balancing AI-generated insights with researcher intuition, and integrating the results into QualCoder, allows for a transparent and replicable process. This approach empowers researchers to conduct efficient, in-depth qualitative data analysis that is both methodologically sound and enhanced by AI-driven capabilities.

The entire 10-step process is summarized in Table 10.3.

Table 10.3. Summary of the 10-step Process to Qualitative Data Analysis in ChatGPT and QualCoder

Step	OBJECTIVE	KEY ACTIONS
1. Clean the interview transcript	Prepare data by cleaning, anonymizing, and formatting transcripts for analysis.	- **Remove Unnecessary Text:** Eliminate filler words, irrelevant phrases, and unnecessary prompts. - **Anonymize Sensitive Information:** Replace personal identifiers with generic labels (e.g., "Participant 1"). - **Structure Data for AI:** Clearly distinguish between interviewer and participant responses.
2. Set up chatgpt for qualitative data analysis	Configure ChatGPT effectively using default or custom GPT models.	- **Choose Your Approach:** Select default GPT-4 for general tasks or build a custom model for specific needs. - **Configure the Model:** Provide specific instructions; fine-tune with relevant data if custom. - **Adjust Settings:** Set temperature to 0.2–0.4 for focused responses; be mindful of token limits.

3. Define themes and develop prompts	Develop a strategy for generating initial codes using deductive or inductive analysis.	- **Deductive Analysis:**
		  - Identify predefined themes from literature or frameworks.
		  - Create prompts targeting these themes.
		  - Refine prompts based on AI responses.
		- **Inductive Analysis:**
		  - Use open-ended prompts to let themes emerge naturally.
		  - Iterate prompts to explore emerging ideas.

4. Create prompts and run coding process	Execute coding by running tailored prompts and saving outputs.	- **Create Clear Prompts:** Align prompts with research objectives. - **Run Prompts for Each Transcript:** Submit transcripts to ChatGPT; save outputs. - **Iterate and Validate:** Refine outputs with follow-up questions. - **Save Outputs for Analysis:** Organize results for later use.

5. Import data into qualcoder and generate a code tree	Transfer ChatGPT results into QualCoder for further analysis.	- **Upload Transcripts:** Import cleaned transcripts into QualCoder. - **Generate Code Tree:** Organize ChatGPT codes hierarchically. - **Transfer Codes:** Manually input codes and associated quotes. - **Review and Adjust:** Refine codes to reflect content accurately. - **Apply Codes Consistently:** Use the code tree across all transcripts.
6. Analyze codes in qualcoder	Finalize coding and analyze themes across the dataset.	- **Conduct Cross-Analysis:** Explore theme variations among participants. - **Generate Reports and Visualizations:** Summarize code frequencies and relationships. - **Interpret Findings:** Use codes and themes to answer research questions.
7. Develop themes and categories	Identify larger patterns and connections in the data.	- **Group Codes into Themes:** Combine related codes into broader categories. - **Use ChatGPT for Suggestions:** Prompt AI to suggest overarching themes. - **Finalize Themes:** Refine and adjust themes within QualCoder.
8. Thematic analysis in chatgpt	Use ChatGPT to perform deeper thematic analysis.	- **Prepare Summarized Data:** Compile codes and key excerpts. - **Develop Advanced Prompts:** Craft prompts to explore relationships and patterns. - **Run Analysis:** Input data and prompts into ChatGPT; review AI-generated themes. - **Refine Themes:** Iteratively ask follow-up questions for depth. - **Document AI Contributions:** Save and organize AI-suggested themes.
9. Validate and cross-check	Ensure coding and themes are rigorous and transparent.	- **Compare AI and Manual Coding:** Check for consistency and accuracy. - **Triangulate Findings:** Validate results with traditional methods or additional data sources.
10. Visualize and analyze in qualcoder	Utilize QualCoder's tools for further analysis and visualization.	- **Generate Code Trees:** Visualize code and theme structures. - **Write Memos:** Document insights and reflections. - **Export Findings:** Prepare data for reports or additional analysis.

Chapter Key Points:

- *Remove filler words, anonymize personal information, and format transcripts to distinguish between speakers.*
- *Choose between default GPT-4 or custom models, provide specific instructions, and adjust settings for focused responses.*
- *Use deductive analysis with predefined themes or inductive analysis to let themes emerge, creating and refining prompts accordingly.*
- *Develop clear prompts aligned with research goals, run them for each transcript, refine outputs, and save results for analysis.*
- *Upload cleaned transcripts to QualCoder, organize codes hierarchically, manually input codes and quotes, and ensure consistent application.*
- *Conduct cross-analysis, generate reports and visualizations, and interpret findings to address research questions.*
- *Group related codes into broader themes, use ChatGPT for suggestions, and refine themes within QualCoder.*
- *Summarize codes and excerpts, create advanced prompts to explore patterns, run analyses, refine themes, and document AI-generated insights.*
- *Compare AI-generated codes with manual coding for consistency, and triangulate findings with other methods or data sources.*
- *Use QualCoder's tools to create code trees and visualizations, document insights with memos, and export findings for reports.*

Part III: Credibility Enhancement and Reporting with AI

Chapter 11.

ENHANCING CREDIBILITY AND RELIABILITY

As AI tools like ChatGPT are integrated into qualitative research, concerns about credibility, bias, and reliability naturally arise. Ensuring rigor in AI-assisted analysis is essential for producing trustworthy and meaningful findings. This chapter focuses on strategies for enhancing the credibility of your research by triangulating AI results with traditional qualitative methods, mitigating AI biases, and using techniques like peer debriefing and member checking to validate the findings. These methods can help you seamlessly blend human judgment with AI-generated insights, ensuring your research remains both innovative and methodologically sound.

TRIANGULATING AI FINDINGS WITH TRADITIONAL METHODS

Triangulation is a well-established strategy in qualitative research that enhances credibility by using multiple data sources, methods, or theories to cross-verify findings. When incorporating AI into the analysis process, triangulation becomes even more crucial, allowing you to balance AI's efficiency with the depth of traditional qualitative techniques.

1. Complementing AI-Generated Codes with Manual Coding

One of the most effective ways to ensure reliability in AI-assisted research is to compare AI-generated codes with manual coding. While ChatGPT can quickly process large datasets and suggest codes based on patterns in the text, it is essential to manually review these codes. Researchers should manually code a subset of the data and then compare the results with the AI-generated codes.

For example, if ChatGPT suggests "workplace stress" as a major theme, manual coding may reveal nuanced aspects such as "workload pressure" or "interpersonal conflicts" that the AI missed. This comparison ensures that the AI does not overlook subtle but significant themes, maintaining the richness of qualitative data. Furthermore, it prevents over-reliance on AI and preserves the researcher's interpretative role.

2. Cross-Checking AI Themes with Theoretical Frameworks

AI can efficiently identify recurring patterns, but the researcher must still validate those themes against theoretical frameworks and the broader research context. Suppose ChatGPT suggests themes like "team collaboration" or "leadership challenges." In that case, you should check these themes against the theoretical frameworks guiding your study, ensuring they align with existing knowledge or concepts.

By overlaying AI-generated codes with theoretical models such as social capital theory or organizational culture frameworks, researchers can deepen their understanding and ensure the analysis reflects meaningful insights. This is particularly valuable when working with interdisciplinary research, as it provides a structured way to align machine-generated patterns with academic rigor.

3. Validating AI Findings Across Multiple Data Sources

In some cases, researchers can apply AI to various datasets (e.g., interviews, focus groups, and field notes) and triangulate the findings across these sources. If ChatGPT identifies the same theme—such as "communication breakdowns"—across different types of data, it strengthens the credibility of the finding. However, differences or inconsistencies between data sources should be explored further, either by revisiting the original data or refining the AI prompts to capture more context-specific insights.

By comparing results across diverse datasets, you ensure that the AI's outputs are not isolated artifacts but robust themes that hold true across different perspectives and contexts.

STRATEGIES FOR MITIGATING AI BIASES

AI models, including ChatGPT, are trained on vast amounts of data from the internet, which means they may reflect inherent biases. These biases can affect the quality of qualitative research by skewing the interpretation of data, particularly when dealing with sensitive or culturally specific information. To maintain credibility, it is crucial to adopt strategies to mitigate AI biases.

1. Critical Oversight and Reflexivity

While ChatGPT can generate themes and codes based on large amounts of text, it lacks the nuanced understanding of context, culture, and emotion that human researchers bring. As a result, AI may reinforce biases present in the data or generate outputs based on skewed assumptions. It is up to the researcher to critically evaluate the AI's outputs, identifying any potential biases and reflexively questioning the AI's "objectivity."

For instance, if the AI disproportionately highlights themes related to "leadership" in interviews with male participants but misses similar themes with female participants, this could signal a gender bias. Recognizing such patterns allows the researcher to intervene and adjust the AI's outputs.

2. Refining Prompts for Balanced Outputs

Prompt engineering is a powerful tool for controlling AI outputs and reducing bias. If initial prompts yield skewed or biased results, iterative refinement can help. For instance, a prompt like "Identify themes related to leadership in this dataset" could be revised to "Identify diverse perspectives on leadership and teamwork across different gender, age, and cultural groups." This encourages the AI to consider a broader set of responses and produce more balanced outputs.

Researchers should continually refine their prompts to encourage diversity in the AI's interpretation of the data, ensuring that underrepresented voices are considered in the analysis.

3. Incorporating Multiple AI Models

Different AI models have distinct strengths and weaknesses, depending on their training data and design. Using multiple models (e.g., ChatGPT alongside Bard or Cohere) to analyze the same dataset can help identify and mitigate biases. If one model emphasizes certain themes while another highlights different aspects, comparing their results can provide a more rounded and less biased interpretation.

This multi-model approach is particularly useful when analyzing sensitive or cross-cultural data, where biases can have a significant impact on the results.

PEER DEBRIEFING AND MEMBER CHECKING WITH AI ASSISTANCE

Peer debriefing and member checking are well-established techniques in qualitative research to ensure the credibility of findings. When combined with AI-generated insights, these techniques can help validate both the machine-generated data and the researcher's interpretation.

1. Peer Debriefing: Cross-Validating AI Findings with Human Expertise

Peer debriefing involves discussing your AI-assisted findings with other researchers or colleagues to ensure the validity of the analysis. For instance, after ChatGPT has helped identify key themes, you can present these to peers for critical evaluation. Colleagues can offer insights on whether the themes align with the data or if there are alternative interpretations that the AI might have missed.

This process helps uncover any blind spots or biases in the AI-generated results, as peers may offer perspectives that differ from the AI's outputs. In practice, this might involve sharing thematic summaries generated by ChatGPT alongside your manual coding for comparison, prompting constructive feedback on both methods.

2. Member Checking: Engaging Participants with AI-Generated Themes

Member checking is the process of validating findings by sharing them with the research participants. AI can enhance this process by providing concise, easily digestible summaries of themes derived from large datasets, making it easier for participants to review and confirm the accuracy of the findings.

For example, ChatGPT can generate summaries of interview themes, which you can then share with participants. If a participant reads the AI summary and feels that it accurately reflects their experiences, this strengthens the credibility of your findings. On the other hand, if participants highlight inaccuracies or missing aspects, you can revise the analysis accordingly.

This method not only builds trust with participants but also ensures that their voices are authentically represented in the final results.

3. Ethical Transparency in AI-Assisted Member Checking

When using AI to assist with member checking, it is essential to be transparent with participants about how the AI was used. Participants should be informed that an AI tool was involved in summarizing or identifying themes, and they should have the opportunity to ask questions about the process. By clearly explaining the role of AI in the analysis, you build transparency and accountability into your research, ensuring that participants are comfortable with how their data is being handled.

Integrating AI into qualitative research offers exciting opportunities for speeding up analysis, but it also introduces new challenges related to credibility and reliability. By triangulating AI-generated insights with traditional qualitative methods, critically addressing biases, and using peer debriefing and member checking, you can ensure that AI enhances—rather than compromises—the rigor of your research.

The key takeaway is that AI should be viewed as a tool to complement, not replace, human judgment. Researchers remain the ultimate interpreters of data, tasked with balancing the speed and efficiency of AI with the depth and complexity of qualitative inquiry. When used responsibly, AI can serve as a powerful partner in enhancing the reliability and credibility of qualitative research, helping researchers navigate complex datasets while preserving the richness of human insight.

Chapter Key Points:

- *Integrating ChatGPT into qualitative research requires strategies to ensure credibility, mitigate bias, and maintain reliability for trustworthy findings.*
- *Triangulating AI-generated codes with manual coding enhances reliability by combining AI efficiency with human interpretive depth.*
- *Cross-checking AI themes against theoretical frameworks ensures that AI outputs align with established research concepts and contexts.*
- *Validating AI findings across multiple data sources strengthens credibility by confirming themes are consistent across different datasets.*
- *Mitigating AI biases involves critical oversight, refining prompts for balanced outputs, and using multiple AI models to achieve diverse perspectives.*
- *Peer debriefing involves discussing AI-assisted findings with colleagues to validate and enhance the accuracy of the analysis.*
- *Member checking engages research participants by sharing AI-generated themes for their validation, ensuring findings accurately reflect their experiences.*
- *Ethical transparency requires informing participants about AI's role in data analysis, maintaining accountability and trust in the research process.*

Chapter 12.

REPORTING AND PRESENTING FINDINGS

Once AI has assisted in the analysis phase of qualitative research, the next challenge is reporting and presenting the findings in a way that is both compelling and true to the data. Integrating AI insights into your reporting requires thoughtful consideration of how to blend AI-generated themes with your interpretations, ensure transparency, and effectively communicate complex themes to a wide range of audiences. In this chapter, we will explore how to craft compelling reports using AI, navigate the ethical considerations of attributing AI's role in the research, and share strategies for presenting complex findings to diverse stakeholders.

CRAFTING COMPELLING REPORTS USING AI INSIGHTS

AI tools like ChatGPT can process and analyze vast amounts of data, but the researcher remains the most important part of the reporting process. Crafting a compelling report involves blending AI-generated insights with your own interpretation to create a narrative that is both insightful and grounded in the data.

1. Organizing AI Insights into a Coherent Narrative

AI models excel at identifying patterns, generating themes, and summarizing large datasets. However, turning these outputs into a coherent and compelling story requires human input. Begin by organizing AI-generated themes into a logical structure that aligns with the objectives of your study. Common strategies include using a **thematic structure**, where each section of your report focuses on a particular theme, or a **case-based**

structure, where you present individual cases followed by thematic analysis.

For example, if ChatGPT identifies "remote work challenges," "employee autonomy," and "work-life balance" as key themes, these could serve as sections in your report. In each section, weave together AI insights with specific participant quotes, manual coding results, and your theoretical framework to create a rich, interpretive narrative.

2. Enhancing Depth with Participant Narratives

While AI can efficiently identify overarching patterns, a compelling report depends on grounding these insights in the raw data—namely, participant quotes, stories, and contextual details. Use the AI-generated themes as the backbone of your report, but rely on direct quotes and case examples to add depth and nuance. This not only enhances the richness of the report but also ensures that the voices of your participants remain at the forefront.

For example, if the AI suggests a theme around "workplace stress," include specific, detailed quotes from participants describing their personal experiences of stress. This humanizes the data and allows the reader to engage with the findings more deeply.

3. Visualizing Data for Greater Clarity

Incorporating **data visualizations**—such as theme maps, word clouds, or network diagrams—can make complex findings more accessible to readers. Many qualitative software tools integrated with AI, like NVivo or QualCoder, offer features that allow you to create visual representations of themes, patterns, and relationships between codes. When these visuals are used alongside narrative descriptions, they provide a dual layer of insight: the story behind the data and a visual summary of its structure.

For example, a theme map showing the relationship between "employee autonomy" and "job satisfaction" could visually highlight the connections that emerged during the analysis, making your findings clearer and more engaging for the reader.

ETHICAL CONSIDERATIONS IN ATTRIBUTING AI CONTRIBUTIONS

As AI becomes more embedded in qualitative research workflows, transparency about its role in the research process is critical. Ethically attributing AI contributions ensures that readers understand how AI was used and that the research remains methodologically rigorous.

1. Documenting AI's Role in the Research Process

Transparency is key when reporting AI-assisted research. You should clearly document how AI tools were used at each stage—whether for coding, identifying themes, or summarizing data. This is particularly important in the **methodology** section of your report, where you detail the processes you followed. Include a description of how ChatGPT or other AI models were integrated into your analysis, alongside any traditional methods you employed.

For example, if ChatGPT was used to generate initial codes, explain how these were reviewed and refined by the researcher. This gives readers a clear picture of the division of labor between AI and human input, ensuring that the final analysis is understood as a collaborative process between AI and the researcher.

2. Balancing AI and Human Attribution

While AI can significantly aid the research process, the interpretation of findings should remain a human responsibility. Researchers must avoid overstating AI's role in generating insights. It is essential to convey that AI assisted in organizing and analyzing data, but the ultimate interpretation—especially the contextualization of themes and patterns—was conducted by the researcher. AI lacks the cultural, emotional, and theoretical understanding that human researchers bring to the analysis.

In the **acknowledgments** or **methodological limitations** section, explicitly attribute what AI contributed to the research. For instance, "ChatGPT assisted in identifying recurring themes and summarizing data, but final coding and thematic development were conducted by the researcher."

3. Addressing Potential Biases and Limitations

It is important to acknowledge the limitations of AI tools in the analysis process, particularly around biases and contextual understanding. AI, especially models like ChatGPT, is trained on vast datasets that may carry inherent biases, and these can subtly influence how themes are identified or interpreted.

Include a **section on limitations** in your report that addresses the potential biases introduced by AI and how you mitigated them. For example, you could describe how AI's outputs were cross-checked with manual coding or how you used peer debriefing to ensure that AI-generated findings aligned with human interpretation.

COMMUNICATING COMPLEX THEMES TO DIVERSE AUDIENCES

One of the key challenges in presenting qualitative research findings is translating complex, often abstract themes into a form that resonates with different audiences. Whether you are sharing your findings with academic peers, policymakers, or practitioners, effective communication is essential.

1. Tailoring the Presentation to Your Audience

Understanding your audience is crucial when presenting AI-assisted qualitative findings. Different audiences may have varying levels of familiarity with AI or qualitative research methods, so the depth and style of your presentation should be adapted accordingly.

- **For academic audiences**, emphasize methodological rigor, including detailed explanations of how AI was integrated and triangulated with traditional methods. Focus on theoretical contributions and how AI enhanced the depth of your analysis.
- **For policymakers or practitioners**, focus on practical implications. How do the themes you identified impact real-world decision-making? Present findings in a clear, concise manner, using visual aids to make complex themes more accessible.
- **For the general public**, simplify the presentation by focusing on relatable narratives and concrete examples. Use plain language and

minimize technical jargon. Visual tools like infographics or case studies can help make the findings more engaging.

2. Breaking Down Complex Themes

AI can uncover intricate and multilayered themes that may be difficult to communicate in a straightforward manner. To make these themes more digestible, break them down into sub-themes or key points. For instance, if your research identifies "remote work challenges" as a broad theme, break this down into specific sub-themes like "communication barriers," "technological difficulties," and "work-life balance issues."

By organizing complex themes into smaller, clearer components, you make the findings easier to understand and more actionable for your audience.

3. Using Storytelling to Enhance Engagement

Storytelling is a powerful tool in qualitative research, especially when communicating complex themes. AI can help identify overarching patterns, but the task of bringing these to life through vivid narratives remains with the researcher. Weave participant stories and direct quotes into your presentation to create a compelling narrative that humanizes the data.

For example, if your AI-assisted analysis identifies a theme of "job satisfaction in remote work," enrich this with quotes from participants describing their personal experiences. This not only enhances the emotional impact of the findings but also makes the themes more relatable to a wider audience.

4. Leveraging Multimedia Tools

In today's digital world, multimedia tools can greatly enhance the presentation of qualitative findings. Integrate video clips, audio recordings, or interactive visuals into your presentations to make the data more dynamic and engaging. For instance, video montages of participant interviews can bring your themes to life in a way that written reports cannot.

Additionally, using interactive dashboards that allow stakeholders to explore the data themselves can offer a deeper, more personalized engagement with the findings.

The process of reporting AI-assisted qualitative research requires careful consideration of how to balance AI insights with human interpretation, ensure transparency, and communicate findings effectively to diverse audiences. While AI tools like ChatGPT can provide valuable assistance in summarizing and structuring data, the researcher's role remains central in interpreting and presenting the results.

By clearly documenting the role of AI in the research process, addressing ethical concerns, and using storytelling, visuals, and tailored presentations, researchers can ensure that their findings are both compelling and credible. Whether communicating to academic peers, policymakers, or the public, the key to successful reporting lies in making complex themes accessible while maintaining the richness and rigor that qualitative research demands.

Chapter Key Points:

- *Blend AI-generated insights with your interpretations to create a compelling, data-driven narrative.*
- *Organize AI-identified themes logically to align with your study's objectives.*
- *Incorporate participant quotes and stories to add depth and maintain data richness.*
- *Use visualizations like theme maps and word clouds to clarify and engage audiences.*
- *Transparently document how AI tools assisted in coding and theme generation.*
- *Balance AI contributions with human interpretation to ensure contextually informed themes.*
- *Mitigate AI biases by critically evaluating outputs and using multiple AI models.*
- *Validate findings through peer debriefing and member checking for accuracy and credibility.*
- *Tailor your presentations to different audiences by adjusting depth and visual aids.*
- *Enhance engagement by using storytelling and multimedia tools to present complex themes clearly.*

ONE LAST THING

If you found this book useful and enjoyed reading it, I would be genuinely grateful if you could take a moment to post a review.

Your support truly makes a difference in reaching others who might benefit from this guide. I take the time to read all the reviews personally, as they allow me to gather your invaluable suggestions and insights, which help me improve this book for future readers.

Thank you sincerely for your support!

It means the world to me and encourages me to continue my work in making research more accessible and understandable for everyone. Your feedback not only motivates me but also fosters a community of learners who can share in the journey of discovery and knowledge.

REFERENCES

1. Jalali MS, Akhavan A. Integrating AI language models in qualitative research: Replicating interview data analysis with ChatGPT. Syst Dyn Rev. 2024 Jul 21;40(3).
2. Şen M, Şen ŞN, Şahin TG. A New Era for Data Analysis in Qualitative Research: ChatGPT! Shanlax International Journal of Education. 2023 Oct 1;11(S1-Oct):1–15.
3. Bijker R, Merkouris SS, Dowling NA, Rodda SN. ChatGPT for Automated Qualitative Research: Content Analysis. J Med Internet Res. 2024 Jul 25;26:e59050.
4. Dengel A, Gehrlein R, Fernes D, Görlich S, Maurer J, Pham HH, et al. Qualitative Research Methods for Large Language Models: Conducting Semi-Structured Interviews with ChatGPT and BARD on Computer Science Education. Informatics. 2023 Oct 12;10(4):78.
5. Brower RL, Jones TB, Osborne-Lampkin L, Hu S, Park-Gaghan TJ. Big Qual: Defining and Debating Qualitative Inquiry for Large Data Sets. Int J Qual Methods. 2019 Jan 1;18.
6. Briganti G. How ChatGPT works: a mini review. European Archives of Oto-Rhino-Laryngology. 2024 Mar 22;281(3):1565–9.
7. Creswell JW. Research Design: Qualitative, Quantitative and Mixed Methods Approaches (4th ed.). Thousand Oaks, CA: Sage.; 2014.
8. Reeves S, Kuper A, Hodges BD. Qualitative research methodologies: ethnography. BMJ. 2008 Aug 7;337(aug07 3):a1020–a1020.
9. Khan SN. Qualitative Research Method: Grounded Theory. International Journal of Business and Management. 2014 Oct 23;9(11).
10. Petticrew M, Rehfuess E, Noyes J, Higgins JPT, Mayhew A, Pantoja T, et al. Synthesizing evidence on complex interventions: How meta-analytical, qualitative, and mixed-method approaches can contribute. J Clin Epidemiol [Internet]. 2013;66(11):1230–43. Available from: http://dx.doi.org/10.1016/j.jclinepi.2013.06.005
11. Weyant E. Research Design: Qualitative, Quantitative, and Mixed Methods Approaches, 5th Edition. Journal of Electronic Resources in Medical Libraries. 2022 Apr 3;19(1–2):54–5.

12. Westbrook L. Qualitative research methods: A review of major stages, data analysis techniques, and quality controls. Information Science Research . 1994;
13. Sutton J, Austin Z. Qualitative Research: Data Collection, Analysis, and Management. Can J Hosp Pharm. 2015 Jun 25;68(3).
14. Novak M, Drummond K, Kumar A. Healthcare professionals' experiences with education in short term medical missions: an inductive thematic analysis. BMC Public Health. 2022 Dec 17;22(1):997.
15. Brownstone LM, Mihas P, M. Butler R, Maman S, Peterson CB, Bulik CM, et al. Lived experiences of subjective binge eating: An inductive thematic analysis. International Journal of Eating Disorders. 2021 Dec 11;54(12):2192–205.
16. Merriman SE, Plant KL, Revell KMA, Stanton NA. What can we learn from Automated Vehicle collisions? A deductive thematic analysis of five Automated Vehicle collisions. Saf Sci. 2021 Sep;141:105320.
17. QualCoder. What is QualCoder [Internet]. 2024 [cited 2024 Nov 2]. Available from: https://qualcoder.wordpress.com/
18. Thomas SC, Neenumol K, Chacko S, Prinu J, Pillai MR, Pisharody S, et al. Feasibility of a nurse-led, mHealth-assisted, and team-based collaborative care model for heart failure care in India: Findings from a multi-stakeholder qualitative study. Wellcome Open Res. 2024 Apr 24;9:219.
19. Namu E, Mberia H. Influence of Narrative Principles on Organizational Change Adoption of Blended Learning in Selected Kenyan Universities. International Journal of Communication and Public Relation. 2024 Aug 16;9(4):15–23.
20. Sushmitha G, Eashwar VMA, Charulatha RJ, Surya BN. Balancing Act: A Qualitative Study on the Dual Nature of Video Game Addiction among medical college students in Chengalpattu district, Tamil Nadu. 2024.
21. Cousins JM, Bereznicki B, Parameswaran Nair N, Webber E, Curtain C. Adverse drug reactions in older people following hospitalisation: a qualitative exploration of general practitioners' perspectives. Int J Clin Pharm. 2024 Oct 19;
22. Rotterdam Exchange Format Initiative. 2024. [cited 2024 Nov 2]. REFI. Available from: https://www.qdasoftware.org/
23. OpenAI. Open AI - Introducing ChatGPT [Internet]. 2023. Available from: Introducing ChatGPT

Made in United States
Troutdale, OR
02/03/2026